LIFE STORY

Many lives, one epic journey

RUPERT BARRINGTON AND MICHAEL GUNTON

WITH

MILES BARTON, IAN GRAY AND TOM HUGH-JONES

FOREWORD BY

DAVID ATTENBOROUGH

BOOKS

Foreword

Dawn in the Kalahari Desert. I was sitting beside a meerkat colony, a dozen or so burrows surrounded by naked earth. The owners were still below, but a researcher who had been studying the group for many months had told me that within ten minutes a baby would almost certainly appear from one of three holes within a yard of me to take its very first steps into the outside world.

Right on time, ten minutes later, an adult female silently crept out, reared up on her hind legs and surveyed the landscape. A few seconds later, a tiny baby, a perfect miniature of the adult, clambered groggily out of the burrow and, while its mother kept watch, took a few steps towards me to inspect my outstretched finger.

It was the first time that this little creature had seen daylight. Its preparations for the great moment had, in fact, been comparatively unhurried. After its mother had mated, it had developed inside her womb for 60–70 days. It was then born in the pitch-blackness of the burrow, where it stayed for 19 more days, suckling its mother's milk and growing in size and strength. Only then was it ready to claim a place in the outside world and face all the dangers that would beset it from then on.

Of course, not all baby mammals arrive in the outside world in such a leisurely way. A baby wildebeest spends eight months in its mother's womb. It then slithers out of her vent on to the open savannah and staggers to its feet. Within an hour it has to be able to walk well enough to follow her as she waders away, searching for grazing. A kangaroo's birth, on the other hand, is either

LITTLE BIG-EARS
A Mongolian long-eared jerboa in the Gobi Desert – a young star of *Life Story* – learning to hunt at night without any parental guidance.

much more leisurely or absurdly swift, according to which moment you decide is its actual birth. If emerging from its mother's body counts as such, then it is born almost unbelievably young for an animal of its size – at a mere 30 days or so. But it is hardly recognizable as a mammal let alone a kangaroo. It is a tiny, naked, worm-like creature with stump-like forelegs and no sign whatever of rear ones. It wriggles up through the fur of its mother's belly and takes refuge in her skinny pouch. It won't even peer out to squint at daylight until almost ten months later – and it won't become independent for another several weeks after that.

Such great variation in the way an animal is born is not restricted to mammals. It occurs throughout the animal kingdom. Fish, of course, lay eggs. But not all do. There are some that give birth to live young, including the guppy, which every tropical-fish-keeper knows well. Nor is live-bearing as recent an evolutionary development as one might suspect. A few years ago, it was discovered, to many people's astonishment, that some of the most ancient of all fish, swimming in the seas of the world 380 million years ago, did the same.

Extend your survey to include insects, and you will discover yet more almost unbelievable ways of entering the world. Greenflies, for example, hatch from eggs in spring. The vast majority of them are female. But if the weather is right and there are suitable plants to provide them with sap, these females dispense with the complications of finding a male and mating. Instead, and entirely by themselves, they produce a tiny infant within their own bodies. And that infant, which will be born within a few days, already has its own tiny infant female developing in its body.

This great variability in the way animals begin their independent lives also applies to the manner in which they deal with the challenges they face at every stage of their lives. Every aspect of a species' behaviour is its particular way of meeting a challenge. There is just one goal – to leave offspring, the next best thing to immortality – and each success leaves an animal that bit closer to the goal. Why should there be such immense variety of behaviour, such immeasurable richness?

One reason concerns the very different anatomies that animals inherit from their distant ancestors. Another is the differing environments that species have to face. But these two simple variables have produced between them an uncountable number of solutions to the challenges of survival through to reproduction. This book and the *Life Story* television programmes survey some of the most fascinating ways of dealing with the trials of life.

Take the art of seduction. A male satin bowerbird spends many years learning the skills to create a bower impressive enough for a female to consider mating with him, while a male Japanese pufferfish's way of seducing a female is to create an architectural sand construction so extraordinary that until recently no one realized it had been made by a fish. By comparison, the male giraffe weevil dispenses with seduction and puts his efforts into guarding a female against the advances of other males as she prepares to deposit her eggs, fertilizing them the moment before they are laid.

With such a vast, indeed almost infinite, number of tactics from which to choose, it is hardly surprising that the programme-makers include examples that few will have heard of and that have never before been recorded for television. And that is why documenting the glory and the splendour of the natural world is a never-ending delight and astonishment, as the following pages prove.

David Attenborough

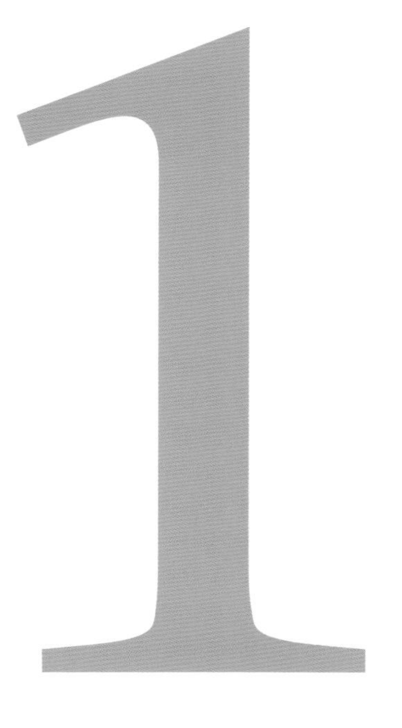

First steps

Birth – the universal moment an animal leaves the egg or womb and begins the journey of life. An animal's ultimate aim is to survive long enough to reproduce, passing on its genetic inheritance. But the likelihood of a newborn finally achieving this goal is slim.

ICE SAFETY
Emperor penguin chicks exploring. While their parents go to sea, they are left on their own in the colony crèche, far enough away from the edge of the ice sheet to be safe from predators.

SIBLING BONDS
Previous page Litter-mates, watching and learning. Their bond will grow stronger as they do. Survival to maturity will depend on the cohesion of the pride as well as their father's ability to defend it against other males.

First steps

For short-lived creatures, this first growing-up stage is all about getting to maturity – fast. For more complex, longer-lived animals, a slower childhood can also be about learning skills that will help them later in life.

Just after birth, animals are typically at their smallest, weakest and most vulnerable. So this first stage of their journey is purely about survival. Luck plays a big part in determining who survives, and each new day is a triumph over the odds. The most effective way to beat those odds is to grow as quickly as possible. So early food intake is usually directly related to long-term success.

Many insect babies are little more than eating machines, processing as much plant material as possible to fuel their transition from pupal stage to adulthood. A grey seal pup will have to survive in the sea, and before it can make the plunge, it must put on as much fat as possible. So it is suckled five times a day on milk that is 60 per cent fat, tripling its size in less than three weeks.

Complex and longer-lived animals usually have few siblings and tend to be brought to maturity by their parents. The mother/infant bond is critical, and these infants rarely survive if orphaned. Even among social primates, where other group members may be willing to care for an orphan, the infant's survival is possible only if it is adopted by a female who has just lost her own baby and can provide milk.

If you depend on your mother for food and protection, you must learn to recognize her. The most basic method, used by waterfowl chicks, is to imprint on the first moving thing seen when they hatch – mother, normally – and then to follow it wherever it goes.

To encourage nurturing behaviour, juveniles often have special features such as big eyes, and they may sport colours that signal to adults that they are youngsters and not rivals. But this doesn't protect against sibling rivalry. Aggression among siblings can sometimes be fatal in animals as diverse as sharks, hyenas and insects. But it is perhaps best documented among young birds. Fighting to the death between chicks in the nest is common and often encouraged by the parents, who can put their effort into raising the chick most likely to survive, especially when food is hard to find.

Other threats close to home include mothers who kill and eat their own young if conditions are particularly unfavourable, so maximizing their chances of being free to breed again when the situation improves. Among some social mammals, dominant males may also kill any offspring that are not their own.

The more sophisticated the species, the more there is to learn to be ready for independence. Animals acquire abilities through observation and practice, and parents or other group members may become tutors. Cats such as cheetahs may deliberately injure a prey animal and then release it in front of their young so that they can develop their hunting techniques. Killer whale mothers spend years helping their young master the art of hunting – for example, how to beach themselves to catch seals. Even domestic hens have been seen to teach their chicks how to select the most nutritious food.

Perhaps the most endearing quality of many young mammals is their playfulness. Having fun is actually essential in developing strength, coordination, timing and social skills. But as life becomes increasingly serious, and youngsters cannot rely on their parents to provide and protect, play tails off. Having completed the end of the beginning of the journey and survived some of the greatest threats, youngsters must now learn to be adults.

SERIOUS MISCHIEF
An inquisitive young langur investigating its reflection in the camera. Its playful curiosity is held in check by its mother.

Strange transformations

The weird and wondrous young of the small majority

STRANGE VIEWS
Top left Snake-mimic hawkmoth caterpillar, Costa Rica, inflating and rearing up its rear end to resemble a snake's head.
Top right Slug moth caterpillar, Costa Rica, pretending to be a seed pod.
Bottom right The pumped up 'head' of a spicebush swallowtail caterpillar protruding over its real head, which has been retracted. The false eyes are to scare off potential predators. If really scared, it can extrude a false forked tongue.
Bottom left Flower-decorated wavy-lined emerald moth caterpillar. When it grows out of its skin, it has to decorate its new one with more chewed-up parts of whatever flower it's eating.

For almost all animals, mortality in infancy is far higher than at any other period of life. Young animals – physically undeveloped and naïve – are poorly equipped to deal with disease and danger. To counter this, there are two strategies. Longer-lived animals such as birds and mammals tend to give birth to few young but take great care of them. Short-lived animals such as insects give birth to large numbers of young, but though they provide limited or no parenting, they have produced so many young that at least some of them will make it to the next stage.

Though most juvenile insects are expendable in this game of numbers, the juvenile life stage is no less significant. Indeed, many species exist far longer as larvae than as adults. The fact that insects have exoskeletons (hard chitin outer casings) rather than bones allows them over a lifetime to easily reinvent their body patterns through the processes of moulting and pupation, which have also resulted in a huge diversity of body forms. They exist as two different – sometimes dramatically different – forms over their lives: as larvae during the young, fast-growing stage and as more sophisticated, sexually mature adults, with structures allowing them to perform tasks such as travelling, fighting and mating.

Caterpillars – the larval stage of butterflies and moths – are effectively eating machines designed to process food into body mass. Rather than going to the expense of developing complex body parts, this relatively simple form allows them to process as much energy as possible into growth, the building blocks of which are then reorganized inside a pupa into the adult form. But being slow-moving and rich in protein, caterpillars need defence strategies to avoiding being eaten.

Common caterpillar adaptations for deterring predators include toxins ingested from plants and associated warning colours, and there are almost endless examples of caterpillars with venomous hairs and barbs. Others use design to scare off attackers. With its cartoonish eyespots, the spicebush swallowtail is a snake mimic with a fake forked tongue that it can shoot out when threatened. Snake-mimic hawkmoth caterpillars inflate their back ends and sway them to create an uncanny impression of

Hatching within a few minutes of each other, they wriggle free and descend a few centimetres on silky lines. While hanging, they expand their bodies to take on the classic mantid appearance

an angry viper's head. Total camouflage is another option, blending in with the plants they live on. One of the most beautiful examples is the looper moth caterpillar that adapts its disguise to match the flowers it is feeding on by decorating itself with their petals.

By contrast, insects such as grasshoppers and cockroaches condense most of their development into the larvae while still in the egg and then hatch as nymphs that have the same overall body plan as adults. Even the nymphs of dragonflies, which live under water, display a similar shape and predatory behaviour to the flying adults they will turn into. Having opted to use the mature stage for the growing part of their lives, such nymphs rely on successive 'instar' moults to recast their outer casings and progressively develop characteristics such as wings and genitalia.

Mantid nymphs, like their adult forms, are predators and hunt from the moment they emerge from their ootheca – the casing that protects 20–300 eggs, depending on the species. Hatching together, they wriggle free and descend a short way on silky lines. While hanging, they expand their bodies to take on the classic mantid appearance. Once their exoskeletons have hardened, the tiny nymphs disperse to find food and hide.

Born just a fraction of the adult size, they must survive five or six moults before reaching maturity. To realize such ambitious growth targets, they will devour whatever they can get their lightning-fast raptorial limbs around. Hunting is partly instinctive, but they must perfect the accuracy of their strike and learn what prey is within their size range, starting with fruit flies and aphids. Many young mantids don't learn fast enough or get lucky enough and die from starvation.

When startled, most young mantids will perform the typical arm-waving, kung-fu-like threat display to make themselves look bigger and scarier. Many are camouflaged in ways similar to their mature forms, but not all. Rather than being white or pink and flower-shaped like the adults, first-instar orchid mantid nymphs are red and black, possibly to mimic a species of assassin bug that has a powerful bite and noxious taste.

Such defensive bluffs, though, are less effective against their own kind. Mantids will attack almost anything they judge they can overwhelm, and that includes even offspring or siblings. So the quicker a young mantis can disperse, the less likely it is to bump into one of its brood. And the quicker it grows, the more likely it is to eat rather than be eaten.

If it survives the first ten days or so, it finds a sturdy spot on which to anchor itself and moult into the next, larger-version instar. As with emergence, some unfortunate individuals don't get it right and perish while moulting. In fact, despite being relatively sophisticated for an insect hatchling, only a fraction of new mantids reach the next instar. But if they do make it to the first moult, it's a major triumph against the odds and a significant step towards ultimate success.

MANTID METAMORPHOSIS
Top left Fully formed orchid mantid nymphs hatching from the protective ootheca in which the eggs were housed.
Bottom left Newly hatched mantid nymphs hanging from silk while their exoskeletons (outer casings) harden.
Top right A young mantid still in red and black warning colours – those used by truly toxic insects – to try to warn off predators.
Bottom right A nearly full-grown orchid mantid, now moulted into its flower disguise, designed to lure insects to its waiting arms.

When life's first step is the biggest

A barnacle gosling leaps first, grows up later

Naïve to the world, physically undeveloped and unable to protect themselves, most animals are at their most vulnerable in the days right after birth. And among the most vulnerable of these are the newborn chicks of ground-nesting birds, including many waterfowl. When their parents are away finding food, they are at the mercy of the local predators.

You might think that migrating north in the summer to nest in the Arctic, as barnacle geese do, might avoid such problems. But there are plenty of predators of eggs and hatchlings even there. And while the comparatively small adult geese can usually defend their goslings from gulls, ravens and skuas, they are no match for Arctic foxes, who specialize in honing in on nesting geese and will take both eggs and chicks. To get away from the Arctic foxes, the barnacle geese lay their eggs on

READY TO JUMP
Two goslings following their mother to the cliff edge as she searches for the best place for them to launch from. The goslings are imprinted on her to such an extent that their need to follow her overrides any self-preservation instinct that might otherwise prevent them jumping off the cliff.

PRECIPITOUS VIEW
Opposite page A barnacle goose family nesting on one of the highest cliff spires. Their next destination – the river and lake – are just visible in the background. The two-day-old chicks will have to plummet to the scree far below, climb over the steep rock and boulder fields and then walk 3km (nearly 2 miles) to the river.

precipitous cliffs that should be too difficult for foxes to scale. The foxes, though, are agile, forcing the geese to find the most vertiginous nesting spots possible, sometimes as high as 200 metres (656 feet) above the ground.

If the parents have selected a good enough site and protected most of their two to six eggs from the cold, some should hatch, and the goslings can usually enjoy a secure introduction to the world. But unlike many other birds, waterfowl and game birds don't feed their young at the nest. Instead the chicks, which are capable of walking shortly after hatching, are expected to eat alongside the adults.

After a day or two of grace – during which the goslings mostly sleep and rely on their yolk sacs to supply the energy for warmth and development – it becomes increasingly urgent for them to descend to the lakes or rivers below. But at two days old they are a long way from being able to fly, and so there's only one way of getting off the cliffs – jumping. It is a leap of faith that the parents have to encourage. They do so by calling, signalling that it's time to go.

Natural self-preservation tells the goslings to stay put, but at the same time instinct produces an even stronger urge: they are imprinted on their mother and find it impossible to resist going where she goes. When the father feels conditions are right, he flies down to perch below the nesting

GOOSE CLIFFS
The glacial valley and goose-nesting cliffs of Orsted Dal in east Greenland. Barnacle geese are gregarious and prefer to nest in concentrated groups. They return to the same nesting ledges on these cliffs, which are used generation after generation.

ledge and calls some more. And when the mother joins him, the chicks can't help trying to catch up with her. Haphazardly, they waddle to the edge, leap into the void and plummet down the cliff face.

They can't fly, but they can slow their descent by flattening their bodies, splaying their webbed feet and pumping their wing stumps. Survival is largely down to whether they can maintain a controlled upright position. Their low weight and fluffy down renders them surprisingly bouncy, and as long as they land belly-first, they can absorb the impact. But if one takes off from a poor spot, gets caught by a gust of wind or hits a protruding rock, it can tumble out of control. Some might survive these mishaps, but many break their necks, sustain other fatal injuries or become stuck in crevasses. Only about two thirds of the breeding colony's goslings survive the fall, and the harsh odds are by no means over. Now the dazed chicks have to try to rejoin their parents. The parents call and search the scree until they're satisfied they've rounded up all the survivors. Then they begin the arduous trek to water, which might be a long distance away. Even though the tiny goslings are running low on energy, they can quickly cover large distances over rough terrain. But now comes the time when they can't avoid their chief predator.

The commotion caused by the jumping goslings attracts the attention of Arctic foxes. They can hear the contact calls between the parents and goslings from far away, and as soon as they spot the parents on the scree, they know goslings will be close by. The foxes launch themselves at the parents, viciously snapping at them. The parents try to fight back, but they have to retreat to

avoid injury. And if the parents are scared away, the foxes dispatch the chicks one by one, eating enough to satisfy their hunger and burying any surplus for the leaner times ahead.

Whether Arctic foxes find the goslings or not depends if they happen to be close enough to hear the goslings' contact calls, but as more and more nests empty, the foxes become more attuned, and fewer broods manage to pass by them.

The cliff-nesting strategy seems an unfeasibly high-risk one for a first step in life, but barnacle geese are long-lived, and a pair will stay faithful to each other and may breed over many years. So as long as every pair eventually raises two chicks to adulthood, the population remains stable. The goslings lucky enough to survive the fall and escape the foxes still have to run the gauntlet of attacks by predatory birds as they make their way to water, and they're not safe from predators until they can fly. But by far the most dangerous period in their lives is now behind them.

GOSLING FATES
Opposite left A gosling falling past a parent. The parents call to encourage their goslings to jump, but they can't help them land.
Opposite right A gosling plummeting head-first. It has to try to right itself, both to slow its descent and ensure it lands on its padded belly.
Above left An Arctic fox digging a hole to cache a gosling. Resident foxes bury most of the glut of goslings as a store for lean times.
Above right A family arriving at the river. Though the goslings are still vulnerable to predators such as gulls and skuas, they can at least escape onto the water.

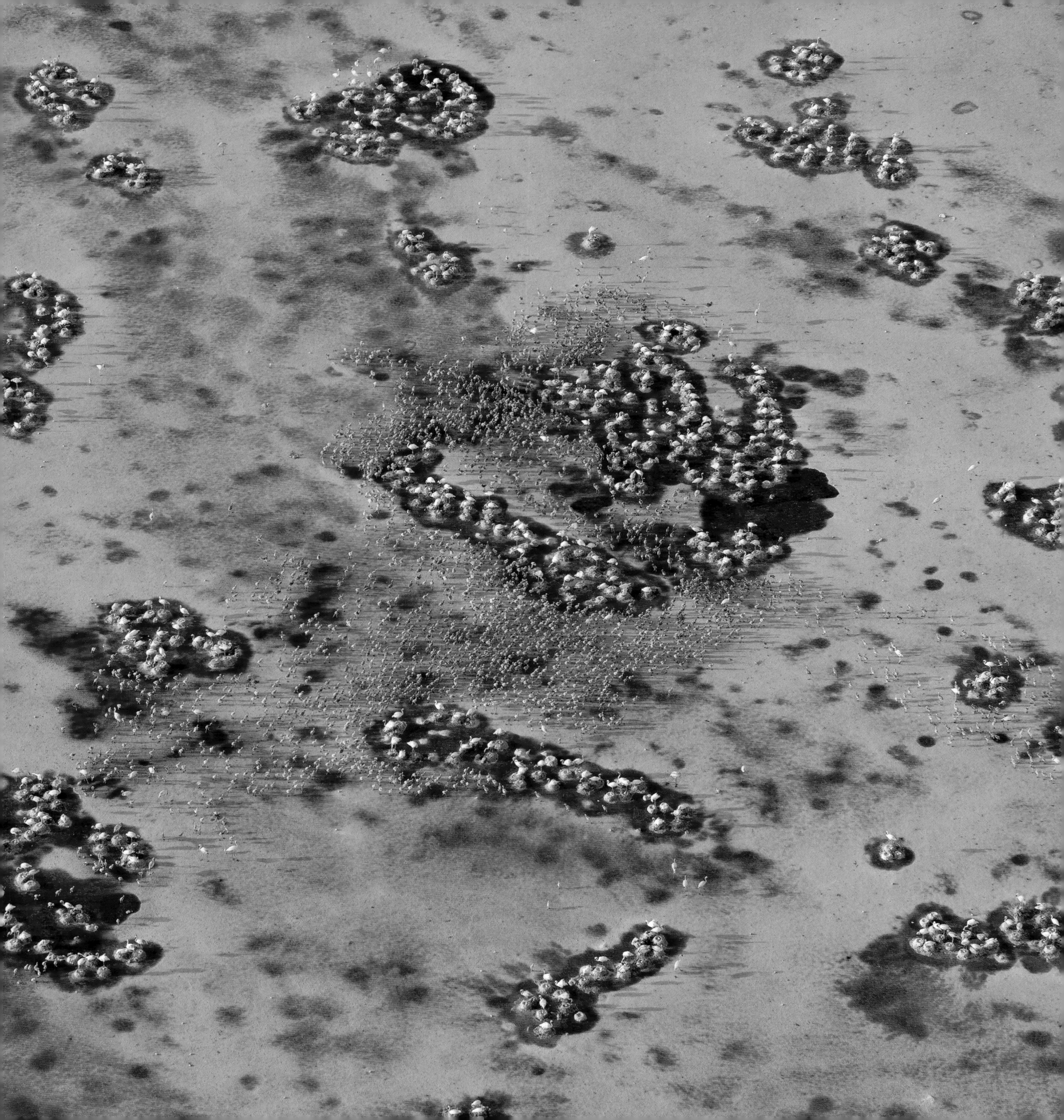

A flamingo's baptism of fire

When you're brought up in hell, there's no problem with predators

NATRON'S NURSERY
An aerial view showing groups of flamingos and their chicks within a larger breeding colony on Lake Natron. Nesting in the middle of the lake provides protection from land predators, which can't cross the caustic water – red from the cyanobacteria that thrive in the alkaline conditions.

Going from the security of an egg or womb to the harsh realities of life outside can be a shock to any newborn, but very few have a shock as extreme as the chicks of lesser flamingos. Tanzania's Lake Natron is the only successful breeding place for East Africa's two million lesser flamingos – roughly three quarters of the world population. It's a very unlikely nursery – a vast, shallow lake that at times is almost as alkaline as ammonia and hostile to almost any kind of life. So the chicks that grow up here face enormous challenges, even though the very conditions that make life difficult also give them their best chance of survival.

After a month of developing inside the egg, a lesser flamingo chick begins to hatch, coaxed along by gentle preening from its parents. Both parents will stay with it during its first few days of life. The chick does little then but raise its head from time to time to be fed and to push with its undeveloped wings. Otherwise it rests on top of the nest cone that its parents have built to keep it above the caustic water and in a slight breeze.

Natron lies close to the equator and gets very little rain, and temperatures regularly reach around 45°C (113°F). So one of the parents' duties is to shade the chick from the scorching sun. They work in relays, switching between umbrella duty and heading off to feed.

Lesser flamingos are highly specialized filter-feeders that eat nothing but spirulina, a cyanobacteria that is one of few life forms thriving in the caustic water and is responsible for creating the red and brown swirling patterns that colour the lake. The parents digest and process it into the 'crop milk' that they feed their chicks. Crop milk is similar to mammalian milk in protein and fat content, though stained blood-red from the spirulina. It is also the chick's only source of water.

The chick remains on the cone until, after four days or so, it is strong enough to climb down and explore its surroundings. As it begins wandering around, soda-encrusted mud collects on its long scaly legs. At first this doesn't matter much, but in the case of some chicks, it can quickly become a problem. Nobody knows why,

MICRO-CRÈCHE
A group of chicks, just a few days old, pecking at insects landing on a nesting cone. Though they are still being fed on their parents' crop milk, they are supplementing their diet with insects and are already associating with other chicks.

OFF TO THE BIG CRÈCHE
A group of chicks, now with curved beaks, moving off to join the big crèche in the open water. Now that they have adult beaks they are able to filter-feed on the same algae that their parents do.

but mud accumulates more on some chicks than others, forming shackles and becoming a burden to the extent of weakening the chicks so they die.

Then there's the unpredictable rainfall. 'Islands' exposed by receding floodwaters reveal fresh, soft mud, and from this the parents can build nesting cones. But sometimes the rains don't stop, which can decimate a colony's breeding attempt. If the water levels rise too high, any unhatched eggs are ruined, and the youngest chicks drown.

On the whole, though, chicks are born and survive in huge numbers. After all, there might be more than 500,000 adults in a nesting colony. But they didn't flock to breed at Natron just to feed there. There are many other, more benign soda lakes in East Africa that are also used for feeding. The reason the flamingos raise their chicks at Natron is that the vast lake is so exposed, treacherous and caustic that the centre is unreachable by predators such as hyenas, jackals and baboons.

But there's no way of preventing aerial attacks. Marabou storks, which regularly fly over the lake, are usually the first to discover the breeding flamingos, and they

become a persistent menace, striding through the colony snapping up as many chicks as possible. The parents do what they can to distract and defend, but though they stand a metre tall, they are fragile birds and are easily overpowered. The storks only relent once they are fully gorged, but by then they may have taken a considerable toll.

The surviving chicks endure for three to five weeks, during which they begin forming 'micro-crèches', spending less time with their parents. Then one day a few of the adult birds start to call the chicks and to lead them away from the muddy colonies towards deeper waters. From large to small, depending on which hatched first, all the chicks from a particular colony join the exodus. The only ones left behind are those too weak to follow – whether through malnutrition, abandonment or mud shackles. Alone in the deserted colonies, they soon perish.

Most do make it into one of the colony's crèches and begin a more nomadic existence: adults alternate child-minding responsibilities, marshalling the chicks around the lakes and leading them sometimes as far as 20km (12 miles) to drink from

Marabou storks, which regularly fly over the lake, are usually the first to discover the breeding flamingos, and they become a persistent menace, striding through the colony snapping up as many chicks as possible

freshwater springs dotted around the lake's edge. Every day the parents return to the crèche to feed their chicks – which, amazingly, they are able to pick out from the crowd and the chicks can recognize their calls.

The chicks are big enough now to be safe from marabou storks, but their need to drink brings them in striking range of the hyenas and jackals that prowl the lakeside and charge into the water to attack the crèches. Eagles also begin attacking both young and old flamingos. Throughout their infancy, many chicks perish, but their crèching strategy ensures safety in numbers. With up to 100,000 in a single group, predators can only account for a small fraction.

The chicks remain together for about 11 weeks, until their growing beaks begin to curve and they can start feeding like adults. Soon after, when their flight feathers appear, they are ready to fledge. Finally, they can escape from Natron and begin flying with the flock to search for feeding opportunities in the soda lakes dotted along the Rift Valley.

They might not return to Natron for years, until they themselves are ready to raise their own young. Then the birds join a mass return to the burning heart of the lake and embark on one of the planet's most spectacular breeding displays, 'dancing' in unison before pairing up and laying their eggs.

CAUSTIC CRÈCHE
A large group of maturing chicks feeding in the open water of Lake Natron. Only a few adults keep watch, but each day the parents visit their chicks to supplement their diet and to lead them to fresh water.

The get-fat-quick diet

Why baby humpbacks have to drink their mothers dry

The rate at which a young animal grows during the early stage of its life can be critical. The bigger and stronger it becomes, the better it will be able to defend itself and compete for food or social status in later life. In the case of humpback whales born off Hawaii, calves must pile on the pounds as fast as possible if they are to survive the long migration from their warm birthing waters to the cold waters of their northern feeding grounds in Alaska. Being among the largest animals on the planet, this requires an intense bodybuilding regime.

The very first thing a calf has to learn to do is suckle under water. It needs to dive below its mother, nudge her body to stimulate her to expose a nipple and then roll its tongue into a straw-like shape to create a seal around the nipple. The thick, rich milk contains 40–50 per cent fat, and a newborn can consume an incredible 45kg (100 pounds) of milk a day. But being an air-breathing mammal, it has to adapt to aquatic life before it can use all these calories for growth.

A young calf is very buoyant and will spend about 85 per cent of its time at or near the surface. It will alternate frequent shallow dives with short bouts of quick breathing, punctuated with periods spent twirling around at the surface while slapping its flippers and tail. Such frenetic exercise helps increase levels of myoglobin – an oxygen-storing protein vital for efficient breath-holding and diving. Though travelling at the surface requires up to six times more energy than swimming under water, the pay-off is long-term efficiency, since vigorous activity builds stamina as well as strength.

After a couple of weeks, a humpback calf enters a more sedate phase that gives both baby and mother rest time. The safe, warm nursery waters provide no food for the mother, and so she is on a tight energy budget. While her calf is growing fat, she's shrinking. She therefore needs to rest or sleep as much as possible to conserve energy. The calf will alternate between diving down 6–8 metres (20–26 feet) to rest beside her and returning to the surface for around two minutes to breathe and swim around in circles.

Now it can swim at a greater depth, it often moves with its mother, staying just above her pectoral fin, and can remain submerged for longer. It also begins

HANGING OUT WITH MOTHER
A baby humpback playing close to the surface while its mother rests. A calf needs to be very active to build up its muscle strength and diving ability. Its mother, on the other hand, must rest as much as possible to conserve energy, as she cannot feed in the warm but unproductive nursery waters.

experimenting with different locomotion styles and breaching. As well as being good exercise, breaching may be how a calf learns to communicate with visual and sonic signals.

Being more relaxed, the calf can use most of the milk it suckles to grow and put on weight in preparation for the 4,830km (3,000-mile) migration to the northern feeding grounds. It needs to be as chubby as possible to endure the voyage and move fast enough to escape the threat of killer whales along the way. Only when it has at least doubled its weight can mother and calf leave for the feeding grounds.

Lack of food isn't the only disadvantage for a mother hanging around the birthing ground. The calves' peak growing period coincides with the arrival of attentive males. Though lactating (milk-producing) females are extremely unlikely to conceive, males will often escort mothers and babies, which can waste critical energy by forcing them to keep on the move. If more than one suitor joins in, the situation can escalate into a full-on 'heat run', where as many as 15 males may pursue a female, bumping and jostling each other for position. The situation can become highly aggressive, and any calf that gets caught up in the fighting will become exhausted and is vulnerable to injury.

Being born early in the season is the best defence. Early calves have more time to grow before breeding activity begins in March and are also able to escape for northern waters sooner. Calves born later spend a greater period of their critical growing time being hassled by male escorts, and so are far less equipped to complete the migration.

LEARNING TO SWIM
A humpback calf swims close to its mother along the coast of the island of Maui, Hawaii. Possibly 2,000 humpback calves are raised in Hawaiian waters every year. Exactly where they are born is a mystery, but the world's highest density of young calves is found off the west coast of Maui.

But, as the *Life Story* team witnessed, the presence of male humpbacks is not always detrimental.

Tiger sharks are common in Hawaii, and though they rarely target healthy calves, they home in on the sick, injured or abandoned ones and are thought to account for more than half the deaths of baby humpbacks in nursery waters. When filming aerials of humpbacks, the team spotted a badly injured calf being harassed by tiger sharks, its mother struggling to keep the sharks at bay. The calf was saved by the arrival of males, which defended the calf by blowing streams of bubbles around it and chasing off the sharks (though sadly the calf was probably too badly injured to have survived the attack).

There appears to be considerable cooperation between the males, and it seems they use fin and fluke slaps either to call more whales for help or to use sound to scare off the sharks – a reminder that there are far more complexities to humpback sociality than we currently understand.

About 80 per cent of calves do make it to the northern feeding grounds. Once there, they can grow even faster, benefiting from the increased milk production from their feeding mothers. By the time a calf is 11 months old, it is 6 times its birth weight and about 8 metres (26 feet) long. It is ready to be weaned and now strong enough to start to fend for itself and slowly begin to enter the adult world.

COPING WITH A CONSORT
A male escort blowing bubbles beneath the mother and calf he is following. Why males pursue lactating females when they are unlikely to be fertile is not fully understood. The significance of bubble-blowing is also not fully known, but males often do it to display strength, dominance and aggression.

The pride of the pride

In the extended lion family, a cub gets care from its mother, tolerance from the aunts and uncles, and security from dad

When animals are raised in families – rather than just produced in quantity and left to luck – it's the quality of family life that's the main influence on each animal's future. A lion cub's long-term survival depends more than anything on the strength and stability of its pride, because in violent lion society, the pride is a cub's only security. But even within the pride life isn't simple.

Lionesses usually have litters of one to four cubs. The newborns, blind and vulnerable, are kept separate from the pride and nursed by their mother in a secluded den, hidden from other predators, as well as buffalo, which often try to kill young cubs. Only at a month and a half are they mobile enough to be introduced to the pride.

Because the pride's females often synchronize their reproduction, all the litters tend to be born and eventually be initiated at about the same time, and growing up with a lot of half brothers and sisters has advantages. The cubs are close in age, which ensures they get a fair share of food and aren't outcompeted by dominant older siblings. Large litters also mean that same-sex littermates will be likely to stick together and support each other if and when they leave the natal pride and attempt to establish themselves in the adult world.

The crèche is at the core of the pride and is a caring cooperative, but only up to a point. Cubs aren't able to suckle indiscriminately, as generally mothers reserve milk solely for their own offspring. The main motivation for communal mothering, it seems, is defence. Cooperating mothers are far more effective at protecting their young and fending off outside males.

Though mothers tend to humour the antics of cubs that aren't their own, whether non-breeding pride members are as tolerant depends on personality. Some of the aunts may be more jealous and aggressive, and subordinate bachelor males can be short-tempered. So the sooner the cubs grasp the pride's social dynamics, the better.

Perhaps a cub's most important relationship is with its father. The pride male will fight to keep his genetic lineage alive and is fundamental to the safety of the cubs. Outside males are always looking for opportunities to overthrow the ruling male or males and

PRIDE DYNAMICS
A male lion snarling at a defensive lioness and her cubs. The relationships between male and female members of a pride can be complicated. The females rely on the males to help defend them and their cubs from external males. But while they are usually submissive to the alpha male, they are often less tolerant of other pride males who might be aggressive towards their cubs. For the cubs, part of growing up is learning about the pride's social dynamics.

FIRST MEETING
A young cub investigating its father, probably for the first time. Cubs don't spend much time with their fathers, and he shows little tolerance of their playful advances, but he will let them join him on a kill – before he allows the females to feed.

conquer the pride, and if they manage that, they will try to kill any cubs from previous males and bring the females back into heat. A new pride male has no interest in being a stepfather, and unless the female escapes with her cubs, infanticide is inevitable.

Childhood prospects improve dramatically if a cub's father can maintain his power. He provides very little in terms of hunting or child-minding, but he will frequently visit his cubs to check they are alive and unthreatened. He also tolerates his cubs feeding alongside him far more than he tolerates females (even though the females may have made the kill, and they usually have, they still have to wait to eat until the male has finished). For cubs, the best father is one who is newly in possession of the pride: he's more likely to be in his physical prime and to endure while the cubs grow up. And as he will have mated with all the females after taking over, the cubs will also reap the benefits of growing up in a large communal litter.

If the pride is stable and the young cubs are able to flourish, behavioural differences between the sexes will begin to emerge. Female cubs generally explore more and are more occupied with hunting-related play. Males often stick closer to their mothers for longer and spend more time play-fighting. The friendship bonds and natural hierarchies that same-sex playmates form are important and long-lasting. Females will stay together and cooperate in the pride, and males very often strike out in coalition as they disperse and attempt to take over another pride.

The cubs that survive childhood go on to perpetuate the blood-and-violence culture that shapes lion society, but most don't make it that far. Altogether, about 80 per cent of young lions will die before the age of two. Some are simply unlucky, but the majority of fatalities are related to some sort of weakness in the family. The pride is everything to a young cub, and the quicker it learns to live within it the better.

BROTHER BONDING
Two maturing male cubs playing. Soon they will have to leave their natal pride and become nomads. The more allies they have, the more likely they are to endure.

HANGING OUT
Overleaf Cubs trying to keep cool by lying up in an acacia tree. Growing up with similar-aged siblings provides the best start in life for cubs.

The meerkat's last lesson

Learning to take the sting from the scorpion

WATCHING AND LEARNING
A young pup watches a carer catch and kill a scorpion, learning all the while how it's done. The anticipation is the chance of being given the meal, minus the sting in the tail.

BEGGING AND PLEADING
A pup practising the art of persuasion. It's begging from a female meerkat, not its mother, hoping that she can be persuaded to provide some milk if nothing else. The more food it gets, the quicker it will grow and the more successful it is likely to be.

Meerkat pups are born into a highly complicated social world. For their first few weeks, they develop underground, looked after by a series of babysitters. By about two weeks, their ears and eyes open, and after three weeks, they emerge, to be greeted by the adult members of the group. From now on the pups must begin to work out how to negotiate meerkat social life.

The first few days in the outside world are tentative. The pups remain close to the burrows and regularly go back underground. Adult group members all continue to help raise the litters of pups that more often than not have been produced only by the dominant female and male. Non-breeding females still produce milk and share the nursing duties. Other members bring back the pups' first solid food to accustom them to a meerkat's varied desert diet.

Meerkats live in arid regions of southern Africa, where food can be scarce and seasonal. Though predominantly insect-eaters, they also feed on spiders, scorpions, snakes, frogs, eggs, plants and, more rarely, small mammals and birds. Discovering what to eat and where to find it is essential to a pup's development. There's an enormous amount to learn, and only a few days after the pups emerge from the den, the helpers take them on their first foraging expedition.

On these trips the group members share the responsibility for feeding and guarding the pups, irrespective of how closely related they are. Using a range of vocalizations, the attentive adults encourage the young to follow them and to observe how they forage. At first the naïve pups are completely unable to find, identify or process their own food. Instead their tactic is to beg. Their relentless pleading calls coax the helpers to surrender the food they've found.

Mastering this art of persuasion is key to a good start. The amount of food a pup manages to acquire dictates its growth rate, which in turn is directly related to long-term survival and breeding success. After a while, though, the pups have to refine their technique by gauging when a helper's generosity is exhausted and it is time to move on.

BABY SNACK
A young pup crunching its way through a scorpion that has been killed for it to eat.

GRADUATE MEAL
This pup has passed the final test – killing a scorpion that has been released for it to practise on. It takes a patient helper to teach a young one how to do it.

As the pups get older, the helpers become less inclined to feed them. Now it is critical for a pup to be able to identify which adults are the most benevolent and which are the most prolific. How good at feeding a helper is depends on many factors including sex, age, hormone levels, foraging ability and temperament.

Adults can vary greatly in their willingness to provide for the pups, but when some give less, the others will compensate. If a pup is able to recognize these trends and concentrate on the most productive individuals, it can dramatically increase its food intake and competitiveness.

The pups are weaned at about eight weeks, and by about twelve weeks they stop being fed by helpers. To prepare them for this time, the helpers try to teach them how to fend for themselves. At first this instruction is purely through example, but as the pups' ability grows, the helpers begin to demonstrate where to search for different types of food and how to process it. Each food item requires a different technique. Larvae have to be dug up, ants licked from the fur, and eggs broken into – and all this takes practice. The patient helpers present novel food items to the pups and then help them work out how to deal with them.

Perhaps the most distinctive lesson is how to tackle scorpions, and it is one of the clearest examples of active teaching in the animal kingdom.

Eating scorpions is a course that's covered in stages, because scorpions are dangerous and good fighters. Their pincers can inflict a nasty nip, and though not fatal to meerkats, a scorpion's sting can be painful. So to introduce the pups to such feisty prey, the helpers present them with scorpions that are progressively less immobilized.

At first the inexperienced pups get scorpions that are nearly dead, so that they can get used to handling and eating them. Once the trainees have passed this test, they are presented with scorpions with their stings disabled (by biting the tail) but their pincers intact. They practise batting at the scorpions with their paws and snapping at them with their teeth while dodging the flailing pincers, but it inevitably takes a few attempts and a few nips to the nose before the skill has been mastered.

Finally, it is time for the ultimate challenge. The tutors dig up and release fully functional scorpions so their students can have a go at disabling and killing them. The first attempts can be epic rounds. The helpers keep watch for predators but are also ready to intervene if a pup doesn't seem to be doing very well. More often than not, though, the pup finally manages to dispatch and eat the scorpion.

It's a rite of passage, and any pup that has made it this far is on its way to becoming self-sufficient and a valuable member of the group.

2
Growing up

Birth, growth, reproduction and death are the fundamental stages of all animals' lives, but a growing-up phase spanning youth to adulthood is only experienced by relatively few. The numerous, less complex and short-lived animals such as insects develop rapidly. Their behaviour is largely instinctive and they don't require a learning period in which to benefit from experience. But longer-lived animals such as birds and mammals usually do have a transition period, when they gain essential life skills.

▶ **STUDENT AND MASTER**
A youngster closely observing an adult brown capuchin using a hammer-rock to crack nuts on an anvil. It's a technique practised by capuchins in Piaui, Brazil, and passed down the generations.

▲ **ADOLESCENT AND MOTHER**
Previous page A young African leopard investigating the camera, watched over by a relaxed mother. Soon after it's a year old, she will become intolerant of her adolescent, forcing it to hunt for itself and become independent of her.

Growing up

In human lives we recognize the period of transition from childhood to maturity as adolescence. It's the time when we become independent from our parents and develop sexually – a period of uncertainty and discovery, when we take new risks. For any mammal or bird in the wild it is invariably a perilous time. It's also when a youngster may be competing with older and vastly more experienced adults of its own species.

The journey to independence varies, but the first step typically involves separation. In primates, this can be drawn-out and painful for both parents and offspring. Birds, on the other hand, tend to have a very short transition to maturity.

The most prolonged parenting of course usually involves the longest period of growing-up. Skills such as finding food, evading danger and social interaction can often only be achieved by trial and error. But nature is unforgiving, and even the simplest of mistakes can be costly. So parental guidance can make all the difference to learners. Also, for social species, it is essential to have learnt the etiquette of the group to avoid being excluded, which can be not only dangerous but also stressful to the point of death. Many youngsters avoid being picked on or thrown out by having colours or characteristics different from those of adults, signalling that they are not yet sexually mature and therefore not a threat.

Once hard times such as migrations, food shortages or weather extremes kick in, the inexperienced perish. In the case of a mammal or a social bird, personality may play a role in how it fares. Conventional wisdom would suggest that the bold and adventurous risk-takers are the ones that succeed, but in the case of mammals, studies show that often it is the more cautious individuals who survive the growing-up stage.

To increase their chances through this transition period, some adolescents form non-breeding groups, where they can compensate for inadequacy with strength in numbers. Bachelor groups, much like youth gangs in human society, are especially important among the young males of social species such as lions, where aggression is a vital factor in commanding territories and winning mates.

Some juveniles in complex cultures benefit from being mentored by older individuals. Young male elephants, for example, often associate with mature bulls, and it has been shown that as well as sharing their knowledge, the presence of bulls suppresses sexual hormone levels in the young males, preventing them from becoming sexually competitive before they've learnt certain basics.

In certain societies – meerkats and African wild dogs, for example – young females may stay within the family structure, learning from the adults but also helping to raise the young, who may well be their genetic kin.

For long-lived animals, the road to independence provides the time to build up strength and to experience competing for territory and mates. Those that have endured this rigorous selection process and are now equipped to compete in the adult world are one step closer to the ultimate goal: reproduction and genetic immortality.

NEARLY ADULT, NEARLY FLYING
A nearly fledged black-footed albatross exercises its wings in preparation for take-off. Its parents abandoned it when its adult plumage started to show, and now it's almost ready to fly. If on its maiden flight it gets past the sharks lurking offshore, it will have a chance of reaching maturity – to return to the island six years later to breed.

The coldest rite of passage

After an easy start, an Arctic fox's life just gets harder and harder

Born during the brief clemency of summer, an Arctic fox pup's first months are relatively easy. Though there can be up to 18 pups in a litter (the largest litter of any mammal carnivore), there are rich summer pickings – the peak of the lemming population, ground-nesting birds and other breeding animals – and so in those first months there is little competition between siblings. But the days of plenty are short-lived. By late August, summer is all but over, migratory animals are heading back to their wintering grounds, and the freeze is setting in. Bleak times loom.

Providing for such a large family quickly becomes an overwhelming burden for the parents, and sibling rivalry escalates. If there isn't enough food to go around, in-fighting becomes fierce, and less dominant pups soon perish. Then, at about 14 or 15 weeks old, the surviving young are forced to leave the family and head out into the freezing

FROZEN PROSPECT
A lone Arctic fox in the frozen landscape of Churchill, Canada, listening for lemmings under the snow. It lives a solitary life and will remain active all winter. But the shortage of food and the extreme cold means it has only a 20 per cent chance of surviving its first winter.

COLD COMFORT
A young fox napping on frozen snow in Canada. Its double layer of fur helps it survive temperatures that can drop below -50°C (-58°F). With such insulation, sleeping out in the cold can be more efficient than expending energy on burrowing.

landscape to begin a solitary, nomadic life. By now, the pups have a snow-white, long-haired pelt. It camouflages them against the icy landscape, and the highly insulating fur enables them to endure extreme cold.

The young foxes survive on stamina and wits, travelling farther in search of food than any other land animal – more than 5,000km (3,000 miles) in five months. The more desperate a fox is, the broader its diet, and it will eat seaweed, berries and even the faeces of other animals. Many develop scavenging strategies, hanging around human dwellings or following polar bears to scavenge from their kills, darting in to steal scraps. It is an uneasy relationship: polar bears can kill and eat an Arctic fox.

Midwinter is the most testing time. Polar bears are hibernating, closing down that scavenging option, and temperatures fall to -50°C (-58°F). The youngsters can tough it out, not eating for days, but just one in three makes it through a normal winter. When pushed to extremes, a young fox's inexperience will count against it. If, though, a fox survives its first year, it has a good chance of making it through its second.

FROZEN FEAST
A young Arctic fox getting the last meat off a bowhead whale jawbone in Alaska. Scavenging is how most foxes survive in winter.

SCRAPPING
A group scavenging from caribou remains. When there's competition for food, the youngsters will lose out to the adults.

BEAR PICKINGS
Overleaf A fox checking for scraps around a sleeping polar bear, Hudson Bay. Foxes that keep their wits about them are usually too fast to be caught.

Sink or fly

Sharks teach albatrosses what their wings are for

TESTING THE AIR
Facing into the wind while paddling in the shallows, a young black-footed albatross gets used to the sensation of lift as it exercises the flight muscles of its huge wings.

EMERGENCY TAKE-OFF
A tiger shark lunges at an unsuspecting youngster that has landed in the shallows after a short maiden flight. Tiger sharks congregate off French Frigate Shoals, Hawaii, every year at the exact time to take advantage of these naïve youngsters.

The passage to adulthood is often the biggest period of change an animal will ever experience. It leaves any comforts of home and care of parents and must learn the skills necessary to survive on its own. For some this is a slow transition. For others it is abrupt, and few animals experience this upheaval more acutely than fledging black-footed albatrosses.

Each February, about 20,000 chicks are born on the French Frigate Shoals, a cluster of remote Hawaiian islands. Brought up as singletons, life is paradise for their first four months. On such isolated islands there are no land predators. So they can spend more than four months lounging in their seaside nests, building up their strength by gorging themselves on a relay of regurgitated fish dinners flown in by devoted parents.

Then one day, around the end of June, without warning, the adults set off out to sea, never to return. For a while the deserted chicks just sit there, bemused by their parents' disappearance. But eventually hunger forces them to take action. With nothing on the land for them to eat, it's time to leave their island sanctuary and venture out over the Pacific Ocean. All of a sudden these sedentary and pampered chicks have to turn into pilots, navigators and hunters.

Flying comes naturally but not always immediately. One by one the chicks instinctively begin to stretch their metre-long wings against the breeze and practise flapping in preparation for their maiden flights. These awkward adolescents are eager to learn but are still uncoordinated and cumbersome. They're not particularly large by albatross standards but still weigh more than 3kg (7 pounds), and getting airborne is the hardest part. First take-off attempts vary in success. A few take to the wing almost effortlessly, but most barely get off the ground before being toppled by the gusty crosswinds that persist over the islands.

Eventually almost all manage to get airborne, but while becoming accustomed to the principles and excitement of flight, many don't travel all that far. A high percentage of birds crash-land in the shallows and must recompose themselves

FLYING THE GAUNTLET
A lucky fledgling manages to take flight over an approaching tiger shark. Those still on the beach seem unaware of the danger. Yet when they take off, most of them will crash-land in the shallows. The lucky ones will manage to fly a distance out to sea before landing, where they are less likely to be attacked.

before a second attempt. This can take a while, and there's little sense of urgency. After all, these birds have never experienced the sea before, and while paddling around in water also takes some getting used to, it is more straightforward than flying, and rafting in the shallows is also more relaxing. But it pays not to hang around.

Every year tiger sharks congregate in these waters at exactly the right time to take advantage of this fledging event. These large, powerful sharks come to prey on the birds from Laysane and Midway, but it is off French Frigate Shoals that their efforts are most concentrated. Dark shadows loom in the turquoise waters, and triangular fins cut through the waves. But the sea-bound albatrosses appear unaware of the danger until, suddenly, a tiger shark surges towards one of them, lifting its head out of the water with jaws full of serrated teeth.

Some sharks hit their target dead on and drag the fledgling under in one swift motion. But most strikes are not so clinical. It's like bobbing for apples, especially as a surfacing shark creates a bow wave that pushes the albatross just out of reach. All too often a desperate struggle begins with the thrashing shark trying to sink its teeth into its prey while the bird flaps around pecking at the snout and eyes of its attacker.

Life hangs in the balance. If the albatross is injured or its feathers have become waterlogged, it usually succumbs to the persistent attacks. But the lucky ones manage to pull off spectacularly narrow escapes, discovering in the process the art of a water

take-off. There can be little better incentive to get airborne than a shark snapping at your heels. With webbed feet furiously paddling and wings vigorously flapping, the terrified young albatrosses finally propel themselves into the air. It's a stark introduction to the realities of life in the adult world.

Much of a fledgling's fate is down to chance, but there's considerable advantage to leaving the nest early. The longer a chick takes to fledge, the more the odds build in the sharks' favour. As more and more albatrosses crash into the sea, the sharks begin refining their techniques, and by the end of fledging season they rarely miss. Yet those adolescent albatrosses still waiting to fly the nest don't seem to learn from observation. The chicks left on the beach witness the gruesome attacks on their fellow fledglings but give no sign that they recognize the danger.

It can seem like a relentless massacre, but only about 10 per cent of the fledglings succumb to shark attack. Far more die each year from the more insidious problem of oceanic plastic pollution. Well-intentioned parents feed their young plastic items that they mistake for food. The indigestible matter collects in their stomachs, and if too much plastic is consumed, they die.

For the chicks that do manage finally to take wing, life will never be the same. The young albatross will wander across the ocean for three years or so without ever landing, only doing so when they return to the islands of their birth to raise a new generation.

SURPRISE ATTACK
A newly fledged albatross recoiling as a tiger shark gapes at it. A shark must judge its strike just right to get its jaws around an albatross bobbing on the surface. The albatross will fight back, pecking at the shark's eyes, and if the attack fails, it has a chance to attempt a water take-off before the shark strikes again.

It's tough being a teenage tiger

Survival not only means learning to hunt but also how to avoid conflict – with other tigers as well as people

TIGER MOTHER AND CUBS
Kankati and two of her three cubs, followed by the *Life Story* team. Identified by her missing eye, which was lost in a fight, she is the most famous female tiger in India's Bandhavgarh National Park.

TEENAGE SPAT
Two of Kankati's cubs play-fighting at seven months old. Play develops hunting and fighting skills and establishes a sibling hierarchy.

The tiger is Asia's top predator, but few tigers survive the battles and endure long enough to reach the position of reigning supreme in a territory. For its first year and a half, a tiger cub enjoys a relatively easy existence, if its mother is a skilful enough hunter to provide for her family and if their father prevails long enough to defend his territory. But when independence calls, life quickly becomes complicated, as the story of Kankati's youngsters proved.

The *Life Story* team followed the fate of three tiger cubs – two sisters and a brother – in Bandhavgarh National Park in central India. Born to Kankati, a first-time mother, who had lost an eye in a territorial fight with another female, the cubs' outlook was unfavourable. But it soon became clear that Kankati's determination to provide for them more than compensated for her impaired sight. The team left them when they were about seven months old, healthy and playful.

Returning to film them a year later, the team found the three cubs were almost full size and just beginning the journey to independence. Perhaps because their mother's disability made hunting for four difficult, the process of separation had started slightly earlier than normal, but all seemed to be going well.

A subadult tiger's journey to full maturity is long. It must hone its hunting skills, cope with ever-increasing competition from its siblings and eventually establish its own territory. Females begin to be in direct competition with their mothers and so tend to start hunting and searching for their own territory earlier than their brothers, who often stick with their mother, relying on her for food to grow as large and powerful as possible so they can defend themselves against other males.

Even in protected areas of jungle that teem with prey, learning to hunt is a challenge. A neighbourhood-watch system is in place. Chital, sambar, wild boar, langurs and peacocks respond to each other's alarm calls, and it's almost impossible for a young tiger to hunt without being rumbled. If it does manage to creep into position, making an ambush kill requires perfect timing. Tigers are capable of impressive short-distance acceleration, but they lack speed or endurance. If they strike too early,

FIRST KILL
The most skilled of the young tigresses with a monkey kill. At first she shared her prey, but then hunger brought conflict among the siblings. To catch deer and boar, they needed to practise their stealth and surprise hunting techniques, but the arrival of the new male put paid to that – and to two of them.

the prey easily outruns them, but if they wait a little too long, their cover is blown. Refining their technique proved a frustrating process for all of Kankati's offspring. But one of the two females progressed faster than her siblings, and at about 19 months she made her first kill.

As long as they have enough reserves and motherly help to fall back on, most young tigers master the art of hunting, but establishing their own territory is less certain. A tigress may allow her female offspring to reside in close proximity, but conflict will arise if they begin competing for food. For male tigers the situation is worse. They require much larger areas that may incorporate up to seven female territories, and a father will only tolerate his maturing sons in his territory for a short time before forcing their dispersal. After this they keep a low profile on the peripheries of other territories.

A young tiger's struggle for space in modern India is made more difficult by the lack of suitable habitat. Outside the handful of reserves, there's little room for tigers. Some survive by hunting livestock, but it's often only a matter of time before they are killed by poachers, farmers or vehicles.

For Kankati's three cubs, the need to move on arrived early, when their father, the area's dominant male, was overthrown by a new male intruder. Once a new male takes over, he has two priorities: to mate with the resident females and to eradicate any cubs he had not fathered. Knowing her cubs were in danger, Kankati left them in an area of thick scrub on the edge of her territory and then lured the new male as far away as possible with the promise of mating.

Though Kankati's strategy worked, the three inexperienced cubs were now having to fend for themselves. The most accomplished female managed to make the occasional kill, but she was becoming increasingly resentful of having to share. Hunger and inequality created tension, and playful fighting soon escalated into more serious conflict. The cubs struggled to survive, getting thinner and more exhausted in the pre-monsoon season, when temperatures soared to 46–47°C (115–116°F).

Then one morning, the park warden discovered a dead tiger, later confirmed to be the female cub that was proving such a promising hunter. The fact that her hindquarters had been part-eaten was compelling evidence of her having been killed by another tiger, and the most likely culprit was the new male.

The crew returned to the UK saddened by the death of the cub they'd followed for so long and to be leaving the two remaining youngsters in a precarious position. They had witnessed the struggle of a top predator as it embarks on the road to independence and power. Two months later, the other female was found dead. Again the new male was the prime suspect. The surviving cub now faces an uncertain future alone.

He is, though, still alive, hunting in an area of his mother's territory. But Kankati now has new cubs, and very soon she is likely to throw her older cub out of her hunting grounds. With the park enclosed by fences, it remains to be seen if he can find enough space to call his own territory without coming into conflict with another male or with people, should he stray outside the park confines.

THE SURVIVOR...
Bala, Kankati's surviving cub. His siblings were both killed, probably by the new male. When tigers are confined within parks, conflict is inevitable.

...AND HIS YOUNG AUNT
Overleaf Bala's half-aunt – another young tiger trying to find a territory. She'll soon come into conflict with people.

Growing up as gangsters

Abandoned by the adults, juvenile Johnny rooks have only got each other

SUMMER PROSPECTS
A striated caracara looks over a vast food resource – the albatross breeding colony on Beauchene, Falkland Islands. In summer, there are eggs and chicks to snatch and dead birds to scavenge. But in winter, when the colony is deserted, caracaras have to hustle a living.

TOUGH LESSON
A juvenile caracara in a submissive pose in front of a pair of adults. Territory-holders will viciously discipline any juveniles that they find trespassing – even their own offspring.

The storm-battered outreaches of the Falkland Islands are home to most of the world's population of striated caracaras, a resourceful raptor known by locals as Johnny rook or flying devil because of its brazen curiosity and mischievous nature. It possesses formidable weaponry, as do all its falcon relatives, but it also displays opportunistic behaviour more typical of the crow family. In fact the striated caracara is considered one of the most intelligent of all the birds of prey.

Surviving around 1,000km (620 miles) farther south than the common caracara, this tough bird relies on its sharp wits to scratch out a living in unforgiving subantarctic conditions. It's an existence that's especially harsh for a youngster embarking on independent life.

As chicks, striated caracaras feast on a glut of rich pickings from the nearby albatross and vast penguin nesting colonies, brought to them by their attentive parents. But as soon as they've fledged, at around five months old, the juveniles are kicked out of the family nest and aggressively prevented from returning. It couldn't happen at a worse moment. This far south, summer is short, and as the dark, drawn-out months of freezing temperatures set in, the breeding seabird colonies disperse.

Lacking in experience, and in direct competition with the larger, more powerful adults, teenage caracaras stand little chance of making it alone through their first winter. So they form gangs and use strength in numbers. Food is scarce, but a mob of young birds become a force, able to muscle its way in on feeding opportunities. The gangs travel from coast to coast, leaving behind a trail of havoc.

These mixed-sex groups can comprise up to 40 individuals, each with inquisitiveness and confidence magnified through pack mentality. Able to adapt their tactics to exploit any situation, they scavenge carrion, attack seals, dig for earthworms, overturn rocks and even fish in tidal pools. This rowdy rabble takes every chance to live up to the reputation of mischievous Johnny rook, hanging around human settlements where they can raid bins, search for scraps and generally make a nuisance of themselves. They also have a penchant for stealing red objects such as clothing,

probably because their excellent eyesight is attuned to the colour of meat. But it is their habit of attacking lambs and weakened sheep on the Falklands that has earned them such a bad reputation among the locals and led to their persecution. Now only about 500 adult pairs survive there, on isolated islands.

Young caracaras are quick to learn, and as their confidence grows, the gangs' activities become more calculated and effective. Though the long, lean winters take a heavy toll, when summer finally returns, the survivors take advantage of millions of seabirds that flock to the Falklands to begin breeding again. The gangs of young caracaras invade the penguin and albatross colonies, bullying adult birds and snatching chicks and eggs and any titbits they can find.

But the highly territorial adults viciously control access to the colonies, even if it means injuring their own offspring. Any juveniles caught trespassing on a territory risk a severe beating. The gang can triumph, though. While the adults are chasing after

Lacking in experience, and in direct competition with the larger, more powerful adults, teenage caracaras stand little chance of making it alone

one or two, the rest can move in to plunder the nests. Though a few may have to take a hit for the team, playing this game of numbers works as a strategy.

Life as a mobster can, however, have its drawbacks. There's little altruism in a community created through necessity. A strong hierarchical structure prevails, with viciously enforced pecking orders and constant power struggles. Longer-standing gang members pick on new recruits and bully weaker individuals on to the periphery. The most bedraggled birds rarely make it, but without the security of the gang, they would perish faster. Despite the improved survival rate afforded by gang culture, only about one in twenty striated caracaras ever reach maturity at five years old.

If a youngster survives four years, it will have moved through the ranks to become one of a dominant few. In its fifth season, it will develop full adult plumage and finally put gang life behind it. Now its aim is to find a lifelong mate, settle down and rear as many young as possible. So its first battle is to secure a territory, and this time the caracara, male or female, will be fighting alone.

AERIAL PROSPECTS
A gang of juveniles soaring over a seabird colony and scanning for hunting or scavenging opportunities. A group may be able to overpower larger animals, even livestock.

THE GANG
A group of juvenile caracaras scavanging on the coast. Such a gang can number 30 – strong enough to plunder the adults' territories. But within it there is bullying and constant power struggles.

How brains are better than bones

A young octopus has to master a repertoire of magic tricks

When they reach early adulthood, many creatures carry on building up their strength and skills so they can effectively compete for a territory or a mate or both. For some, this period is relatively brief, but for others it is the longest and most critical phase in their lives.

Octopus adolescence begins when a hatchling is large enough to catch animals bigger than the tiny plankton drifting at the surface of the sea. It sinks to the sea floor, and from now on, the majority of its life will be spent intensively body-building until it reaches the right size for mating.

Most octopuses have relatively short life expectancies and reproduce only once. In fact, it is the act of mating that triggers the onset of death in almost all of them – females stop eating and survive just long enough to brood their eggs, while males slowly deteriorate and die a month or two after. With just one shot at passing on their genes, achieving prime breeding condition is essential. Whether it's the need to fight off rivals or to produce more eggs, bigger is better.

Thanks to their extraordinarily fast metabolisms, octopuses can convert up to 60 per cent of what they eat into body mass and increase their weight by as much as 5 per cent a day. Some species even achieve a 30-fold weight gain in just a few months. To do this, an octopus has to spend a great deal of time hunting crabs and other crustaceans, which means being out and about in daylight. But with no bones or armour, an octopus is a soft target and a rich source of protein for many other sea creatures. So a growing octopus faces a stressful existence, being on an almost permanent hunt for food while constantly on the lookout for predators. The solution is camouflage, aided by its intelligence, flexibility and incredible colour-changing skin.

All octopuses are masters of disguise, able to change shape, colour and even texture almost instantly to blend into the background, whether coral, rock, sand or seaweed. The day octopus is certainly an accomplished disappearance artist, but sometimes it adopts a more brazen approach. It performs a display of colours, stripes

STILT WALKING
A young coconut octopus walking stiff-legged on its two strongest tentacles. When travelling long distances over the sea floor, it learns to walk rather than swim – for an octopus, a more efficient means of locomotion.

and spots so startling that it confuses or unnerves a predator long enough to escape.

Other octopuses take this 'hiding in plain sight' principle to extremes. Sporting high-visibility black and white stripes, the wonderpus octopus swirls and loops its long arms in a mesmeric, almost psychedelic gait, which leaves potential attackers baffled, revealing one advantage of being boneless.

When threatened, the algae octopus can 'crinkle' up six arms into a bunch-of-algae pose and then use its other two limbs to tiptoe away. But the mimic octopus does something even more remarkable. It contorts its stripy body into a repertoire of shapes and poses, which appear to be impersonations of poisonous and aggressive sea creatures such as sea snakes and lionfish. It's not clear if it adapts its mimicry in response to different threats, but it can certainly change its behaviour according to the type of predator.

If caught in a really tricky situation, an octopus may resort to squirting an inky smokescreen in an enemy's face and then making a jet-propelled getaway. But some octopuses use a truly extreme measure. They sever an arm and leave the wriggling, disembodied tentacle to distract the predator long enough for an escape – better to lose a limb than its life.

Perhaps the most astounding strategy is that of the coconut octopus. Before setting off to hunt, it seeks out shells or coconut halves discarded by humans, preferably a pair. Holding the shells, it blows jets of water from its siphon to clear them out. Then it stacks them inside each other, clutches them under its mantle and 'stilt walks' off on two stiffened arms.

When it finally spots prey, the coconut octopus abandons its shells and switches into hunting mode. But if danger approaches, it grabs the stashed shells and uses its nimble suckered arms to pull them together and

form a spherical fortress around its soft body. If it is unlucky enough to have only one shell, it just hides underneath it.

Usually a coconut octopus remains stationary while hiding, but occasionally it will shift its weight inside the shells in order to roll its way out of trouble. Eventually it pops an eye out of its shelter to check that danger has passed, and if the coast is clear, it gathers up the shells and walks off on two legs.

This remarkable behaviour is considered to be the only true example of tool use in invertebrates. Rather than just using objects to hide inside, the coconut octopus is selecting, manipulating and storing suitable shells.

It's through their remarkable intelligence and unique adaptations that growing octopuses manage to exist as both vulnerable prey and voracious predators. Maturity is determined by the size at which, as a male, he can both win over a female and guard her from other males, and as a female, where she is large enough to produce a huge clutch of eggs and guard them – without feeding – until they hatch and the babies are ready to leave the den and start the growing process.

SHELL CRAFT
Opposite left A coconut octopus, using a jet of water from its siphon to expel sand from a newly selected, well-fitting pair of coconut shells.
Opposite right Stilt-walking while carrying the new shells. It will take the protective home wherever it goes.
Above left Retiring within its home to eat a crab. When hunting, the shells are used for concealment.
Above right Feeling threatened, the octopus uses its suckers to pull the shells around it for defence, and it changes colour to aid the disguise.

A PhD in bowerology

It takes six years of hard work to get your qualification in mate-attraction

YOUNG BLUE-EYES
A juvenile male satin bowerbird investigating the work of a master-builder. It is almost identical to a female, which may be why it's tolerated by the male bower-builder. Both sexes have blue eyes, suffused with lilac-pink, but a male doesn't acquire the black plumage and yellow beak until it's about five years old.

FINAL STICK-WORK
An imposing bower 'stage' for singing and strutting, with much blue window-dressing for impressing the females. Once, only flowers and feathers were available as decoration, but now plastic has taken over.

In the case of a young male bowerbird, acting like an adult means far more than simply fending for oneself. He must also have mastered the cultural rules of his species. In fact, he must spend all of his adolescence perfecting an elaborate art form.

The males of the 20 different species of bowerbird all build decorative bowers, mainly to impress females. Sexual suitability is judged largely on creative skills. The more accomplished the artwork, the greater the male's mating success. But while the potential and drive to create is inherited, becoming a master bower-builder is only possible through learned behaviour, an apprenticeship that may take up to six years. Older bowerbirds that have mastered their trade become mentors, allowing the young birds to study and imitate their techniques and practise alongside them.

This fascinating relationship is best seen among the satin bowerbirds of eastern Australia. Their mating season lasts from October to January, and the males spend many of the preceding months preparing. From as early as June, a master-builder begins choosing and collecting materials from the eucalypt and rainforest habitat so he can create a beautiful bower in time for the big day. He begins by clearing a circular area on the forest floor. Then he starts to build two arched walls of twigs about 35cm (13.8 inches) high and forming a 45cm-long (17.7-inch) 'avenue' – curiously always on a north-south axis – which leads to the bower. His apprentice pays attention throughout.

The relationship between master and apprentice can be incredibly strong, with both parties remaining loyal for many years. But for others the arrangement is far looser, and the young males may even visit a number of accomplished bower-builders across the forest to obtain style tips. In his first year the student may just watch, perhaps picking up the occasional twig. The learning process is dependent on the individual – some juveniles become proficient quite quickly, others take far longer and a few never really improve, failing ever to produce convincing bowers or to attract the right kind of attention. This may be because, as well as learning, there is an instinctive ability involved in both construction and décor, and some birds simply don't have it. By the third or fourth season, the apprentice – if he has been paying attention –

STUDENT
A young male in female plumage inspecting the bower of an adult male in the Atherton Tablelands of Queensland, Australia. It will spend about seven years perfecting its bower technique, acquiring skills by trying to copy the efforts of mature males.

MASTER
A male adds a stick to the same bower (see left). Built far from any human settlement, he has used natural objects to decorate it and will tend it throughout the breeding season.

should be able to construct a bower, albeit a scruffy one. It's at this point that he's ready to take the next step: learning to decorate it. But working out what to use as decoration, where to find it and how to display it around the bower takes time.

Each bowerbird species has its own taste in décor. As well as using yellow sticks and white snail shells, satin bowerbirds adore bright blue decorations. Finding natural blue objects such as feathers and berries was once a difficult task in the Australian bush, but now bottle tops, straws, toys and the like provide an ample source. This abundance has allowed the male birds to let their creative vision run riot. What used to be subtle compositions of muted blues and yellows are now vivid plastic palaces built from litter dropped by people who frequent their forest homes. If anything, the artistic bar has been raised, and the young males have even more to live up to. Inadvertently, the apprentices also end up helping to decorate their mentors' bowers. As an apprentice diligently searches for suitable decorations for his bower, the mentor hops over to it and steals the best items. Returning apprentices often appear slightly bemused but never seem to object to their mentors' thieving ways.

Aside from demonstrating one of the most accomplished artistic feats in the animal kingdom, these natural sculptures double as a stage on which the male has to

give a star performance. Once he has a female's attention, the male begins a bizarre dance, strutting on the spot while presenting the female with prized decorative items in his beak and accentuating the seductive impact of his moves with a vocal whirring noise. But such a virtuoso performance takes practice, and it will be a long time before a young male has the chance to perform in front of an audience.

Until his seventh season, when he finally gains the blue-black satin lustre to his feathers, a young male sports an olive-green plumage, like that of a female – so similar, in fact, that he seems to stimulate the mentor into display mode, dancing to the apprentice as if he was a female. Such ambiguous encounters allow the master to rehearse and the juvenile, clearly rapt by the show, to study and learn the moves.

Useful though apprentices may be, mentors are only prepared to put up with them for so long. As the breeding season approaches and the females begin taking a real interest, the apprentice either becomes a nuisance or, depending on his age, competition. It is only a matter of time before the master destroys any practice bowers and plunders the materials. The apprentices put up with such harsh treatment, seeking tuition over successive building seasons until, after six years, they become artists in their own right and begin entering their own creations into the competition to win a mate.

As an apprentice diligently searches for suitable decorations for his bower, the mentor hops over to it and steals the best items

The non-stop hummingbird

When you leave the nest, life becomes a never-ending whirr of wings

AEROBATIC REFUELLING
A male booted racket-tail hummingbird feeding from an orchid in cloud forest on the western slope of the Ecuadorian Andes. It has to sip nectar thousands of times a day to fuel its hyperactive lifestyle. But finding flowers with enough nectar is not easy, as a flower delays replenishing its nectar supply to make sure that any hummingbird that has drunk from it moves on to a new flower and disperses the pollen it is carrying.

For a young hummingbird, the pressure is on to grow up fast, attain full size and fend for itself. But on flying the nest, life is still full of challenges, and few have a more demanding adulthood than a booted racket-tail hummingbird. Even among hummingbirds, it's on the small side. A male weighs just 3.5g (0.1oz), and half its 10cm (3.9 inches) length is accounted for by its two tail feathers. But though the racket-tail is tiny, it lives its life at an impossibly fast pace.

Every morning, it wakes up hungry and takes off in search of food in the Andes tropical cloud forest where it lives. All hummingbirds are accomplished aviators, able to fly forwards and backwards with incredible speed and manoeuvrability, and smaller species such as the racket-tail are, relative to size, the fastest propelled creatures on the planet. Their wings can beat up to 80 times a second. But this consumes a lot of energy. With the exception of insects, hummingbirds have the fastest metabolisms of any animals and require a prodigious amount of high-grade aviation fuel.

Hummingbird fuel is sugar-rich nectar, but it doesn't come free. Plants require the hummingbirds to transport their pollen, and each flower provides just enough nectar at any one time to carry the bird to the next flower. A racket-tail may have to visit more than 1,000 flowers a day to consume enough nectar (many times its body weight) to stay alive. If it doesn't feed, it will die of starvation within 24 hours. Essentially it is a slave to the flowers, existing on a tight energy budget with little room for error or rest.

As the sun rises and the forest warms, the racket-tail expends some precious reserves hunting for protein, mainly in the form of tiny flies. It takes quite a few of these to satisfy dietary requirements, but the racket-tail's aerial ability and split-second reactions make it an efficient hunter. But it, too, is hunted. Vipers and other snakes dangle motionless next to flowers, waiting for hummingbirds. Hawks and falcons patrol the hummingbird's favoured feeding sites, agile enough to snatch them on the wing. A booted racket-tail therefore remains tuned to the alarm calls of other birds.

By midday the tropical sun reaches maximum intensity, and if the racket-tail has fed well it can relax. But not for long. Downpours typically begin in the heat of the day.

As the rain falls, so does the temperature, and a hummingbird is forced to increase its feeding rate. Even in storms, when raindrops can be almost as big as a hummingbird's head, the racket-tail has to keep looking for food. It can dodge some raindrops, but if the rain persists, it eventually gets wet and burns even more energy keeping warm.

On top of all this, the booted racket-tail has rivals to contend with. It lives in the most hummingbird-rich region of the Americas, where a small area of forest can contain 40 species. Many of these compete for the same flowers, adopting different tactics to ensure they get enough nectar. Larger hummingbirds are often territorial, aggressively guarding a flower-rich area. Some take advantage of more dispersed flowers by memorizing the most efficient route and repeating it on a schedule that allows enough time between visits for each flower to replenish its nectar supply. Others have developed beaks shaped to feed on the nectar offerings of specific flowers that others can't access. Remarkably, many of the smallest hummingbirds operate under the radar of the larger species by mimicking the appearance and sound of a bee in flight.

But what the booted racket-tail sometimes does is hustle for food. It will 'buzz' other hummingbirds, flying at them with noisy wing beats to scare them off for long enough to snatch a quick feed. Such daring raids are risky – with their spear-like beaks, hummingbirds can inflict serious, sometimes fatal, injuries on each other. Another tactic is to wait until two larger species are battling over a flower before darting in for a sip of its own.

The racket-tail's superior agility and acceleration usually allows it to manoeuvre out of trouble, but not when up against its own kind. Booted racket-tails are to some extent gregarious and will even gang up on the bigger species, but when nectar is in short supply, they become less tolerant of each other. Most disputes are settled through ritualized threat displays, when they hover in front of each other, splaying their forked tails and puffing out their leg pompoms. But if neither backs down, a fight is inevitable, and the duelling pair lock claws and stab at each other until one eventually gives up.

Altogether, the racket-tail has a lot to contend with in a day. But as the sun sets, the temperature drops, and the racket-tail has to feed even more frequently. On top of this it has to build up energy reserves before dark to avoid starvation over night. This is the last push – a frantic rush hour when both feeding and fighting reach maximum intensity. Then finally, as night falls, the racket-tail can roost, exhausted.

After finding a thin perch, out of the reach of nocturnal predators, it fluffs up its feathers, closes its eyes and shuts down. By dropping its heart rate to 50 beats per minute, it reduces its metabolic rate to a minimum. For 12 hours the little racket-tail remains in this energy-saving torpor, but as soon as the morning light filters through the canopy it's time to wake up and face another frantic day.

Booted racket-tails are to some extent gregarious and will even gang up on the bigger species, but when nectar is in short supply, they become less tolerant of each other

NON-STOP PROBLEMS
Top, left and right An acrobatic booted racket-tail hummingbird dodging attacks by African honeybees – bees with stings (unlike native bees), which compete for nectar. A sting from one could be fatal.
Bottom left A male clashes with a female booted racket-tail at a drinking spot. When a lot of hummingbirds have gathered to feed, small species may nip in and sneak a drink while the bigger ones are chasing off rivals.
Bottom right Rival males splaying their tails and puffing out their white 'boots' as they fly at each other. If neither backs down, they will begin stabbing with their beaks and grappling. They may even lock claws, fall down and wrestle on the ground.

Home

All creatures need a place to live, somewhere to call home. And for young animals leaving their parents, it's vital that they quickly find shelter from the vagaries of a dangerous world and give themselves a chance of a future. For some, security can come as much from the company an animal keeps as where it chooses to live. For others the key to staying alive is all about occupying territory or creating a permanent dwelling.

NEST UNDER COVER
A great grey owl acts as a duvet cover for her owlets in a Norwegian wood. The owlets will move out of their birch-tree nest-hole at around four weeks old but will continue to be fed by their parents and won't leave home properly until the autumn.

NIGHT MONKEY DAY HOME
Previous page A pair of night monkeys in their day shelter – a rainforest tree hollow, Ecuador.

Home

One way young animals can avoid the problems and dangers of finding their own place is to stay at home with their parents or social group. But stay-at-home youngsters have to earn their keep and may even have to sacrifice their own chances of breeding. In social species, often only one gender is allowed to remain – females in the case of lions and elephants, forming multi-generational prides or herds, working as a group and often sharing child care. In other societies – chimpanzees, for example – it's the males that tend to stick together and the younger females who find a new home with a new group. In either case it avoids inbreeding.

For albatrosses and penguins, home is where they choose to lay their eggs – for the rest of their time they are travellers. Tigers and polar bears wander over huge home ranges that they share with others of their kind, whereas weaver ants and European robins defend a small patch against all comers. Some animals have permanent, house-like homes, dug out underground or built above ground, others take over abandoned homes, and snails, corals and tortoises make homes out of parts of their bodies.

Home can be just for a day, a season or a lifetime. Beavers may occupy the same dam for generations, spanning centuries. But whatever an animal's requirements, its home provides one or more of the most immediate necessities: protection from the environment – cold, rain, heat; protection from predators by means of location or structure; or the provision of resources, particularly food and water, though it may also provide other facilities – a sunning site, for example, needed by reptiles.

The quality of a home can have a huge impact on an animal's chances of survival and ultimately of breeding. But the problem many young animals face is a competitive housing market. The search can be long and dangerous, and the best-quality properties are inevitably occupied. So adolescents may have to settle for something less than ideal. Also, getting a foot on the property ladder may be just the start of their trials. An animal that builds its home from scratch may have to practise for years, as do many nest weavers and even some burrowers. Beavers start with innate know-how, but it may take years to develop the skill to build and maintain the perfect lodge.

As a new property increases in value, it inevitably attracts the interest of others. Some – the symbionts – will give something in exchange for residence. But others are squatters or even have eviction on their minds. So residents must be prepared to defend their investment, whether it be a shell, tunnel or swathe of grassland.

The decision to stand firm and fight or back down and move depends on weighing up benefits and costs. Violent border disputes or more subtle defences, such as signalling ownership through scent or calls, all have costs. A bird singing for hours on an exposed twig to advertise its presence risks being caught by a predator. Scent signals can be expensive, too: urine marking by territorial male house mice, for example, quickly leads to loss of weight.

A significant reason for defending your home is that it can be crucial in attracting a mate. A potential partner is attracted to a territory for the same reasons that its owner was – availability of food and protection from predators, meaning it's a suitable site to rear young. Males with ideal territories may even attract more than one female.

So, for young animals, finding and keeping hold of a property is on many levels an important step in their life journey and crucial if they are to have any chance of success in the game of survival.

LIVING HOME
A pink anemonefish among the tentacles of its anemone host in New Guinea. The little fish never strays far from its permanent home, protected from predators by the stinging tentacles, itself immune to their toxins.

Three weeks old and thrown out of home

A young pika has to both house-hunt and harvest

Pikas may look like small hamsters, complete with big incisor teeth and long whiskers, but they are relatives of rabbits and hares, not rodents, and they are extreme survivalists. They make their homes on some of North America's highest mountains, among the steep boulder fields that collect beneath precipitous cliff-faces.

One of the best places to see American pikas in action is on a vast scree slope known as the Rock Glacier, which spreads beneath Mount Rae in the southern Alberta Rockies in Canada. Here, the huge numbers of jumbled rocks are home to countless pikas, even though the little animals aren't that easy to see.

Though they don't hibernate, they spend most of the long winter in snug dens among the boulders, buried beneath a blanket of snow and living off the food supplies they have stored during the short summer. Caching enough food to last the winter is a challenge, especially for young pikas. Born early in the spring, often when the snow is still deep, pika-lets are on a fast track to independence. Weaned in just three weeks, they are then ejected from the mother's territory to fend for themselves.

It's critical that each one finds a place to settle down as soon as it can, which isn't always easy. Pikas can live for several years, and dominant animals will inevitably occupy most of the good spots. But the combined attrition of winter and predators inevitably creates a few vacant lots each year, and that's what the young pikas are looking for. They have little time to house-hunt. They need to establish a base as soon as possible so that when the last of the snow clears in early summer and the meadows flower, they are ready to begin their frantic preparations for winter.

Having harvested huge amounts of vegetation, they cram it into larders in rock crevices and beneath overhangs, where it is protected from the elements, is aired and can dry out. Hay piles can be enormous. Pikas are high-speed mowers, dashing back and forth between meadow and hay pile, making as many as 14,000 round-trips during the short summer. In a good summer they can store as much as 350 days' worth of forage – more than enough to get them through winter, even with natural wastage.

MY HOME, GO AWAY
A pika defends its territorial boundary by calling from a high point. Its neighbours will recognize its squeaky call and the scent-marks it uses to reinforce its ownership. A youngster looking for its own patch to claim must make do with a home on the outskirts.

These collecting forays take them away from the protection of their home base, and all the to-ing and fro-ing is a risky business. Weasels, coyotes, pine martens and many raptors frequent these slopes, all after a pika.

On their expeditions, the pikas tend to ignore the lush grasses in preference for alpine flowers, biting off stems as close to the ground as possible and struggling back with great mouthfuls of mixed alpine salad.

It's not just plants that are stored. Like their rabbit relatives, pikas can produce two types of solid waste: large, soft, green pellets and small, dry, brown ones. The green pellets still contain nutrients, and so pikas will often collect and store them in their hay piles to be eaten later – a kind of long-life snack.

To keep all this produce fresh enough to eat for several months, pikas turn the contents regularly to slow the rotting process. They are also fussy about which plants they store in their larders. Perversely, many of the plants, such as alpine avens, are loaded with chemicals that make them bitter and difficult to digest. Yet the pikas can't get enough of them, because their chemicals inhibit microbial decomposition, acting as a hay-pile preservative. And during the autumn and winter, these chemicals slowly degrade to the point where the pika can eat them without ill effects.

Not surprisingly, pikas defend their homes and larders ferociously. When not gathering, they stand guard on prominent rocks, barking out warnings to their neighbours and chasing any that intrude too far into their home territories. Those with well-stocked hay piles need to be especially wary – because some pikas are not above stealing from their neighbours. Once a pika has discovered stealing, it may become a repeat offender, turning to this strategy year after year. It simply keeps watch until the owner of a hay pile is busy out in the meadow and then nips in and steals whatever it can carry. But thievery also carries its own risk. If a thief is caught in the act, the owner may react violently.

All pikas, though, are also under a much greater threat: climate change. Rising temperatures are shrinking the size of alpine meadows and reducing the snowpacks that protect them in the winter. Also these high-altitude specialists can't tolerate prolonged high summer temperatures. So the hardships ahead are considerable.

▲ **HAYSTORE**
A pika adds rosebay willowherb to the pile, which is left to dry in the wind and sun through the summer. Once winter threatens, the pika will drag the huge haystack into crevices deep within the boulder fields. Should this larder not last until spring, the pika will burrow through the snow in search of low-growing cushion plants and even lichens.

▶ **LARDER MOUTHFUL**
Making hay while the sun shines. An experienced pika in a prime territory can easily collect enough hay – mainly flowering plants – to stuff several larders. A youngster may not be so lucky.

Waiting for a lifelong lift

Remoras: the permanent passengers

REMORA, SUCKER-SIDE UP
A grounded remora, awaiting a new lift, its sucker fin ready to stick it head-first. Though capable of swimming on its own, it doesn't have a swim bladder, has trouble manoeuvring over distances and prefers free rides.

HITCH-HIKERS' HANG-OUT
Daytime resting spot for a shoal of remoras – a fishing-boat wreck. The marks on the side of the hull show where algal growth has been removed by the suckers of previous hangers-on. When the right transport – a shark, a turtle or even a whale – swims by, they will detach by sliding forward and reattach to the new ride.

Choosing the right home can be one of life's most important (and potentially expensive) decisions. And selecting the best one can be a particularly ticklish problem for a young remora fish.

A remora's idea of settling down is to invest in a mobile home that it can attach itself to. And in the tropical waters around Fiji, young remoras are spoilt for choice. The constant comings and goings at the reef, all day and all through the seasons, provides these choosy fish with the perfect estate-agent's window.

Whether it's schools of visiting bull sharks or migrating manta rays, passing turtles or resident reef sharks, the homeless remoras can peruse such mobile homes at leisure. At Beqa, bull sharks are one of the favourite options, and so when one passes, several remoras may race from cover in competition for these hot properties.

A remora sticks itself to its chosen home using a specialized suction disc on the top of its flattened head. This is a modified dorsal fin – a remarkable and complex structure. The lip of fleshy tissue around its perimeter creates a seal against the host's skin, and an oval disk in the middle, made up of rows of slats or plates known as lamellae, creates the bond between the remora and its new home. Once attached, the remora can slide backwards to increase the pressure of the suction, or wriggle forwards to release its grip.

The disc first begins to show when a young remora is only a centimetre long, and by the time it reaches about three centimetres, the disc is fully formed and the remora can start its career as a hitch-hiker. As it grows, it seeks out larger and larger homes. The suction disc can stick to all sorts of surfaces including the shells of sea turtles, the skin of manta rays, dolphins and dugongs, neoprene wetsuits and even the hulls of ships. Indeed, the name remora is the Latin word for delay, because the weight and drag of clinging remoras was believed to slow down ships.

As a hitch-hiker, a remora can travel far and wide while expending little energy, and it may also gain some protection from potential predators, especially if it has attached itself to a large shark. And being attached to some of the oceans' most prolific hunters means there is never any shortage of food. While remoras usually cling to their hosts' torsos, others are found on or near the gills or even inside the mouths of their hosts. Some even lurk near the rear 'vents' of their hosts to collect waste. They also move over the surface of their hosts, feeding on the bacteria and parasites, so providing a cleaning service, which could be why many of these hangers-on are tolerated.

This attachment is far from permanent. A remora will happily break the attachment seal long enough to scavenge for scraps of food dropped by its host in open water and then chase after and reattach to its host. They also have little loyalty and will take up with another, possibly better host should one happen to be within reach. And while the remoras may provide their hosts with a travelling cleaner service, the relationship can be more one-sided.

The drawbacks for the host may include reduced swimming efficiency, infections, skin abrasions and sores created by the constant suction attachments, and the overall discomfort caused by having to put up with more than 20 of these fish scurrying around on their bodies. But for a remora, once it finds a good ride, hitch-hiking is the most efficient way to satisfy their basic needs.

WHALE SHARK RIDERS
A great ride for a remora – a whale shark, the world's largest fish, with lots of room for hitch-hikers. The remoras' streamlined shape and flattened heads make them less of a hindrance to their hosts – unless, of course, their numbers build up and weight and drag become a problem.

A home so safe it's inescapable

Mountain goats get stuck for the winter on the highest peaks

Some animals choose to make their homes in seemingly unpromising places. Mountain goats live on the near-vertical cliffs surrounding some of the remotest peaks in North America. One of their strongholds is in Glacier National Park in Montana. They seem perfectly at home in this perpendicular world, moving sure-footedly across slopes scoured by avalanches and violent storms, balancing on narrow, icy ledges and scraping a living from what little vegetation there is.

To survive here, a mountain goat has specialized kit. To aid walking on ledges its hooves have spongy, skid-resistant pads and two flexible toes that can spread apart for balance or come together to grip a knob of rock. It also has a dense fur coat, with an outer layer of long guard hairs overlying an insulating layer of wool that traps air.

These goats choose a home in the heights to avoid predators, and as winter bites, rather than head down into the more sheltered valleys, they often climb onto the highest, most exposed ridges. Here, the wind keeps the ground relatively snow-free, and so they don't expend huge amounts of energy trying to excavate vegetation beneath the snow. It's a risky strategy because the deep snow that surrounds these wind-cleared islands can trap the goats for months. So as winter progresses, they slowly eat themselves out of food, and their condition begins to deteriorate. By the end of the winter, all are suffering, but it's the youngest goats that fare the worst.

As kids, they were under the care of their mothers. Adult females rank highest in the social order and vigorously defend their offspring. Their short horns can inflict serious wounds, and so a lot of the fighting is ritualized, sparring and hooking at each other's rumps. But in the spring, when the females give birth again, the youngsters are suddenly a nuisance, and their mothers drive them away.

Once out of the family, they lose all status and drop to the bottom of the pecking order. Being smaller makes travelling and digging through deep snow more exhausting, they lose heat quicker and they have limited access to food. Half of all kids are thought to die before the end of their first winter.

LIFE ON THE LEDGE
Mothers and kids on the tricky way down from a salt lick. In mountain goat society, females outrank males, kids rank the same as their mothers and yearlings are at the bottom of the pile. Top-ranking mothers will get to use the best spots on the salt lick and acquire the best sleeping and feeding areas.

SUMMER PICKINGS
Mother and kid, already free of much of their winter fur, snacking on vegetation on the mountain. The yearlings get pushed to the edge of any feeding areas.

As the spring thaw takes hold, the goats are released from their mountaintop prisons and begin to move down to the lower meadows and the promise of fresh grass and alpine flowers, rich in moisture and nutrients. For the young goats in particular it's one of the most perilous journeys of the year. As they move down from the high cliffs they become potential prey for a whole range of predators, including grizzly bears, pumas, wolves and eagles. Mothers get together and descend with their young kids in tow, following ancient trails. But the yearlings must make this hazardous journey alone, unless lucky enough to hook up with a group of their peers.

As they descend, their thick winter coats become a liability, and they are in danger of overheating. So they shed their fur in great hanks and use any remaining snowfields along their route as rest stops. Reaching the meadows, they stop to graze, but then they move even farther down the valley – after something just as valuable as forage: salt. In the Flathead River valley they find it where the river has carved away the rock to expose a grey clay containing the minerals the goats crave. Calcium, potassium, sodium and magnesium found in the lick may help replace the elements goats typically lose from their bones during the deprivations of winter. The minerals may also act as a digestive aid, helping to absorb nutrients from the summer grasses while preventing diarrhoea and gastro-intestinal disease.

But to reach the lick, the goats have to navigate avalanche chutes, streams overflowing with meltwater and dense forests. For many, the final barrier is the Flathead River itself. For the kids and yearlings especially, it's a testing crossing, but if they make it safely across, the lick offers a refuge. Here they can return to the vertical world they are built for. More than 200 individuals may visit this one site in a single summer.

On these cliffs, the hierarchy of the high tops is re-established. The yearlings are forced into the marginal areas, just as they were on the winter ridges and the spring meadows, and adult females and their kids dominate the prime spots. Once they have spent a few days nibbling away at the lick and restored any mineral deficiencies, they are ready to run the gauntlet of the river and the forest again and make it back to the safety of the high cliffs.

DANGER ON THE DESCENT
The Flathead River crossing – a danger spot en route to the salt lick, especially for yearlings without a mother to lead the way.

THE GREAT LICK
Overleaf Part of a large herd of goats climbing a steep slope above the Flathead River to mine the salts essential for their health. Mothers and kids take the prime slots, yearlings get pushed to the margins.

When home is on the range, the pups are left in care

Why the pack depends on the services of adolescent babysitters

The Liuwa Plains in western Zambia – vast, flat and largely featureless – are where the Sausage Tree clan roams. This large pack of African wild dogs is nomadic, trekking over more than 1,000km^2 (386 square miles), making their home anew each day. Fluctuating dry and wet seasons and a resulting fluctuating supply of migrating prey animals are what force the pack to range widely. The wild dogs also need to keep one step ahead of their most dangerous neighbours – packs of marauding hyenas.

Young African wild dogs stay with their birth pack for several years, serving an extended apprenticeship as they learn how to survive in this testing territory. But membership of such a tight-knit society means earning your keep, and for the youngsters, that means playing a subordinate role to the alpha pair that lead the pack. Their help is especially important when the alpha female gives birth, and the arrival of pups forces the pack to set up a temporary home.

GETTING ACQUAINTED WITH DAD
Just out of the den, very young pups are greeting their father – the alpha male and one of the pack's main hunters. While they are small and need a home for protection they remain at the den, guarded by a babysitter.

DIGESTION SIESTA
Opposite page Pups asleep outside their temporary den. They will have feasted on food regurgitated by the returning hunters.

Settling down even on a temporary basis leaves the dogs vulnerable and robs them of their main hunting advantage, their mobility. The birth den is usually a shallow burrow deep in a dense thorn thicket. The pups – as many as 20 in a litter – remain under ground for their first weeks, and when they emerge they are still too small to follow the hunt. So while the pack fans out across the plains in search of a meal, the increasingly boisterous pups need supervision, which is when the apprentices may come into their own.

While the pack is away, one or more of the younger members, rather than the mother, may stay behind to babysit. It's a heavy responsibility. As well as stopping the young pups from wandering off and getting into trouble, the babysitters protect them from predators such as snakes and the hyenas. Bushfires can sweep across the plains, and once the wet season really gets going, even drowning in the den is a possible threat. At the first sign of trouble, a quick bark brings the pups bundling back to the den. But if the danger is urgent – perhaps the risk of a predator discovering the den – the babysitter may make the risky decision to shift the pups to another den.

Forcing babysitting duties on a subordinate brings other benefits to the pack. The babysitter gains valuable parenting experience, and the alpha female – probably the most experienced hunter in the pack – is able to start hunting again, which is crucial now the pack has many mouths to feed.

But how to bring food back to the den when a kill is made a long distance away? The solution once again depends on the pack working together.

After a successful kill – in Liuwa that's often a wildebeest or lechwe antelope – all members of the pack gorge themselves, filling their bellies with meat, and then they trek back to the home den to regurgitate food for the pups and to reward the babysitter. The pups are

dependent on this regurgitated meat for their first two to three months.

As soon as the pups are strong enough to trek, the clan abandons the temporary home and resumes its wanderings, with the frisky youngsters in tow. At this stage the pups are still too young to keep up once a hunt develops, and so they are parked in cover while the adults pursue their prey, and they are then escorted to the carcass by one of the subordinates. It will be several months before they are tough enough to follow the hunting party and feed directly at the kill, but mobility does mean they can keep one step ahead of the hyenas.

In recent years, this nomadic lifestyle with its need for huge areas of open range has led to a dramatic decline in wild dog numbers. Once they roamed most of sub-Saharan Africa in their hundreds of thousands. Today there are thought to be fewer than 5,000 left. They are victims of drastic habitat loss, which has reduced their food supply and put them increasingly at odds with the lions and hyenas that are their natural enemies.

A MEAL OF A WILDEBEEST
Left to right Assessing the prey, harrying the herd and then singling out a weak member. Though a wild dog is too small to kill a wildebeest on its own, the formidable hunting machine of a pack can. As many as 20 dogs may take part in a kill.

BABYSITTER AND HIS CHARGES
Overleaf A litter of pups at the den, watched over by their young minder. They will stay here until they are 10–12 weeks old and are strong enough to become nomads and follow the pack. A litter can be up to 20 pups, though 10 is more usual, but most won't survive. The African wild dog is classified as endangered, with a population that's still falling.

Hermits on the housing ladder

Crabs queue up for their ideal shell

A few miles off the coast of Belize in the Caribbean lies a palm-fringed coral island no bigger than a football pitch. Carrie Bow Cay is home to hundreds of terrestrial hermit crabs, ranging in size from a fingernail to a cricket ball. Late in the afternoon, the crabs emerge from the shadows of the sparse vegetation to patrol the beaches, picking through the strand-line debris, looking for tasty morsels. But occasionally they stumble upon something far more valuable – a newly abandoned sea-snail shell.

A hermit crab uses a shell as a mobile home, carrying it around on its back and holding it in place with its strong hind legs. The shell protects the crab's soft body parts and provides an armoured retreat when the crab is threatened by predators such as seabirds.

The shells the crabs like best are those of the West Indian top snail, though they will make do with those of clams or scallops or even pieces of driftwood, glass or plastic bottles. They need to upgrade their homes as they outgrow old or damaged ones. But on Carrie Bow, they face a serious problem: there is an acute 'housing' shortage, and a new shell washing ashore is therefore a very valuable commodity.

The first lucky crab to stumble upon such treasure will quickly inspect the prospective new home and then slip out of its old shell to try on the new one for size. If the shell fits, the crab moves in and moves on. But if the shell is too big, a remarkable chain of events ensues.

The crab will stand by its precious find, sometimes for hours, waiting for another hermit crab to find and possibly claim the shell. This 'waiting game' is a high-risk strategy because hanging around on the beach leaves the crab exposed to potential predators and dehydration if their shell is damaged. But on these crowded beaches, it's not long before another crab does arrive. The new visitor will assess the shell to see if it is the right size, and if the fit isn't good enough, it too will linger nearby. Over the course of a few hours, 20 or more 'wrong-sized' crabs might end up hanging out by the empty shell. But they don't just stand around idly; they jostle, arranging themselves in descending size order, forming a line from largest to smallest.

SHELL QUEUE
A line-up of hermit crabs on the beach. Opportunities to upgrade to a larger shell are so few and far between that the crabs will patiently queue for hours in the hope of trading their outgrown shell for a new one.

MOVE 1: NEW HOME
A hermit crab about to abandon a top shell for a larger home. The slipper shell is being checked for cracks and its thickness measured to make sure it will provide protection from both buffeting and predators.

Sometimes the crabs will form several lines, each radiating out from the empty shell, creating a starfish effect. When enough 'crab power' has amassed, the lines of crabs begin a tug-of-war to retain control of the empty shell.

As the lines grow and the squabbles escalate, the smaller crabs will sometimes dart back and forth from the end of one line to the next, trying to predict which one is most likely to capture the prize – just as supermarket shoppers might switch from one cashier's line to the next, looking for the shortest wait.

Eventually, a crab that's a perfect fit for the shell will turn up. As soon as it abandons its own shell and takes possession of the empty one, the second crab in line

abandons *its* old shell and grabs the newly abandoned larger one. This swapping cascades down the line as each crab vacates its old shell and claims the next available one. By the time this swap-fest is over, even the smallest crab at the end of the successful line will have moved into a bigger home, leaving its own tiny shell abandoned on the sand.

This is known as a sequential vacancy chain, and after all the hours of waiting, it's usually completed in just a few frenetic seconds. It's possible that the hermit crabs may have evolved sophisticated social behaviours to make these vacancy chains run more efficiently, especially in the recruitment of other crabs to the cause. Perhaps they release chemical signals or call or display in some way to attract nearby crabs and so initiate a chain.

The term sequential vacancy chain was originally coined by social scientists to describe the ways that people trade resources such as apartments. And it's now believed that quite a few animals use vacancy chains, too, including anemonefish, lobsters, octopuses and some birds, where the swaps occur in a relatively short space of time.

As for those crabs that chose the wrong queue, when another shiny new shell washes ashore, the whole complex process will begin again.

MOVE 2: CHECKING FOR SIZE
Checking out the abandoned top shell to see if it is big enough to move into. Crabs take their time, using their antennae to size up a potential new home. Repeated viewings and even a little interior DIY may also be necessary.

MOVE 3: SWAPPING OVER
Moving from a whelk shell into an abandoned slipper shell. The actual swap-over happens in seconds because of the risk of exposing the crab's delicate abdomen to waiting predators.

A mating weaver has to know his knots

If the male's nest is not good enough, the female just tears it up

FRAME-WORK
Sitting within the framework, as if on a trapeze, a male vitelline masked weaverbird weaves and knots, the diameter of the nest determined by his reach. Different lengths and widths of grass are used for different stages of construction.

HALF-HITCH, LEFT TO RIGHT
The male masked weaver in the process of weaving a new nest from fresh strips of grass, threading them left to right. To make a strand flexible enough to knot and shape, he passes it back and forth through his bill.

Among the most accomplished of home-makers are the weaver birds. There are more than a hundred different species, found mainly in Africa and Asia. All weave elaborate hanging baskets using hundreds or thousands of pieces of vegetation or, in the case of the communal nests of sociable weavers, hundreds of thousands. It's the males who undertake the work, and whether or not a male ends up with a mate often depends on his mastery of the craft – whether a female believes the home is suitable and safe enough for her eggs and the resulting family of fledglings.

In southern Africa, nest construction coincides with the start of the rainy season, guaranteeing a flush of grass for weaving and an abundance of food for the chicks. Gathering grass is an art in itself. Partially cutting through the base of a blade, the weaver holds the cut end and flies up, so peeling away a long, narrow strip. Inexperienced birds may repeatedly shred their chosen strand or bite off too much so they can't tear it away. Only with practice do they become proficient – and they need to be: one nest may require a thousand such fibrous strands. But this is only the first challenge.

A favoured site for a nest is usually a fragile, outermost branch of a thorn-covered savannah tree, sometimes overhanging water. This gives the nest some protection from tree-climbing predators such as snakes but is not the ideal foundation for construction.

Using his feet and beak, the weaver first ties a long strip of grass to the twig. The knot is crucial – from it everything else hangs. Novice weavers may make many attempts before the knot holds, and if it is not tied properly, it can unravel later, dumping the half-built structure onto the ground or into the water. The bird then begins to weave more strips together to create a ring, rather like a trapeze suspended beneath the branch. Then, standing within the swing and always facing in the same direction, he begins to weave a spherical basket, the diameter of which is determined by the reach of the bird.

Individuals vary their technique from one nest to the next, some building from left to right, others from right to left. Their stitching techniques would be familiar to any seamstress, and the knots include spiral bindings, half-hitches and slip-knots. While they are at work, they are assessed by prospective mates. If a male knows he is being

observed, he will work even harder. Eventually, the roof, chamber and entrance tunnel are bound together in a tight-fitting, watertight mesh. As the fresh grass dries it shrinks, tightening the weave and strengthening the structure. The entrance size must be large enough for the weaver to enter but small enough to thwart nest-raiders such as snakes, hawks and cuckoos. To show off the final construction, males also indulge in a bit of landscaping, clipping the leaves off surrounding branches.

To announce an 'open house', a male perches beside it, sometimes hanging upside down from the entrance hole and frantically fluttering his wings – an invitation to females for a viewing. A messy nest is quickly rejected. The female will also assess its strength and age. They have no time for nests that have taken so long to build that the grass has died away – an indication of the male's inexperience. To try to hide their ineptitude, slow-building males will continually add new grass to the outside to make the nest look fresher. But females are seldom fooled, and an unacceptable nest is quickly pulled

Then, standing within the swing and always facing in the same direction, he begins to weave a spherical basket.

WEAVER VILLAGE
An acacia tree festooned with the nests of masked weavers, some hanging from the old nests of much larger birds. Communal nesting offers the females a choice of master-weaver mates as well as safety in numbers.

RENOVATION
A Southeast Asian baya weaver brings fresh palm fibres as the finishing touches to his nest, which has taken at least 16 days to weave. An interested female will poke and pull it to test its construction and then thoroughly inspect the egg chamber.

apart, and the male has to start again. The position of the nest on the tree is another deciding factor. The best sites are on the windward side, and centrally placed nests are favoured because they are potentially the most buffered from predators.

A young, unsuccessful builder will often make many nests over the course of the mating season – some males have been recorded rebuilding nests more than 25 times. But with each new build, speed and technique improve. So in one way, the choosy females are improving the males' future chances of success.

Once a female is suitably impressed and has mated, she will take over the home-making duties and complete the interior decorating, lining the nest with soft grass and feathers. This frees up the male to start building another nest in the hope of attracting another female. It may even be on the same branch, so he can defend both nests from other males intent on ripping them apart. A talented male masked weaver bird may mate with several females in a breeding season – the reward for his home-building skills.

Even the babies get stuck in

When weaver ants weave, no one is idle

Many ant species make their homes in trees, but in the rainforests of Australia and Asia, weaver ants have taken the arboreal way of life to new heights. Rather than depending on natural cavities for shelter, as most tree-dwelling ants do, weaver ants weave leaves together to create some of the most complex structures found in nature.

This ability to sew their own home gives the ants a huge advantage over any competitors, because if the colony needs to expand, the ants can simply enlarge their nests by incorporating more leaves into an extension, or start a new one next door.

The construction of any single 'pod' requires some very sophisticated collaboration between many individual ants. First, workers need to bring two leaves together by pulling with their powerful jaws. If the distance between the leaves is greater than the stretch of a single worker, a second worker climbs over the first.

STRETCH AND PULL
Weaver ants repairing a leaf nest. A team of workers are pulling together the edges of the two leaves until they are close enough to be glued with silk.

THE NEW POD
An extension to the growing weaver ant colony. Additional leaf pods on the borders of the colony include barracks full of workers ready to die in defence of queen and colony.

If they still can't reach, others join the chain until the leaf-edge finally comes within range of the leading ant's jaws.

Once this contact has been made, others swarm across the bridge, repeating these acrobatics until a living network connects the two leaf edges. Then these labourers begin to pull, slowly drawing the leaf-edges together. As the distance is closed, other workers use their jaws to pinch together the seam between the leaves (like staples holding together a sheaf of papers).

More workers now start to arrive at the building site carrying larvae that are about to pupate and metamorphose into adults. To achieve this transformation, the larvae will spin themselves protective silk cocoons. The builder ants take advantage of this, making the larvae donate some of their valuable silk to the colony before being taken back to continue their development.

The larvae-carriers position themselves along the seam between the two leaves. Then each ant taps the head of the larva with its antennae to encourage it to start extruding silk from its salivary glands. The worker swings the larva from side to side, stretching a strand of silk back and forth across the join until a network of threads binds the leaves together. The whole process is repeated with more leaves until a football-sized nest chamber has been formed.

A well-established colony may contain half a million ants living in many such nests. This metropolis may extend across several trees – a complex of districts and suburbs connected by busy commuter routes.

Living in one of the central nests, a single queen rules her empire. Every day she lays hundreds of eggs, which are then carried to the outlying nests to be cared for and fed as they develop into larvae. Two kinds of workers are produced in these nurseries: small minors that stay inside the nests and care for the queen's brood, and larger, tougher majors built for tasks outside.

The queen can live for many years safe inside her chamber, but the average worker usually lasts only a few months. As a major worker ages it takes on ever more risky tasks until in its old age it ends up patrolling the dangerous outer borders of the territory. One of the old ants' tasks is to keep the growing colony provisioned, and they are such efficient hunters that they will systematically strip the home trees of food, creating a virtual desert where few other creatures can survive. Chinese farmers noticed this phenomenon 1,700 years ago and began placing nests in their orchards to help protect the fruit

from pests, making weaver ants perhaps the oldest known form of biological control.

This scorched-earth approach means expansion of the colony is an ever-present necessity. And since all weaver colonies follow this same trajectory, they inevitably come into contact with each other. Then it's war. If a patrol encounters an unidentified intruder, the ants immediately attack, spraying formic acid from their abdomens and biting the invader with their powerful jaws. If the threat is great, some of the defenders fetch help, laying scent trails along the way to guide thousands of reinforcements back to the threat.

But weaver ants have a more caring side, too. Some gangs of workers tend flocks of scale insects and other sap-sucking bugs, guarding them from enemies and in return gathering the droplets of sugar-rich honeydew that the bugs excrete when feeding. This much-valued commodity is carried back to the nest and shared.

They have also developed a relationship with some kinds of caterpillars, which live in their nests unmolested, protected by chemicals that integrate them within the colony. The ants will even carry the caterpillars to grazing sites and watch over them while they eat. In return, the caterpillars secrete honeydew.

But successful colonies also attract unwelcome squatters. One is a small jumping spider that mimics an ant, both in appearance and behaviour, enabling it to move with impunity through the colony, grabbing defenceless larvae whenever it pleases. It's a lifestyle that provides both a ready-made home and a never-empty larder.

But such intruders are minor inconveniences for an empire than can dominate several rainforest trees and number tens of thousands of workers.

MAKING AND DEFENDING HOME
Opposite left Home-extension builders, stitching leaves together by using larvae as silk-extruding sewing machines.
Opposite right Once the pod has been stitched, workers will move the overflow larvae to their new nursery.
Above left Any ants from a neighbouring colony who try to gain access will be fought by the patrolling major workers.
Above right But an ant-mimic spider can move in undetected, to feed off babies.

4

Power

Few animals operate in isolation. Individual lives impact and interweave. It's therefore vital for an individual to know when to fight and when to back down, when to go it alone, when to cooperate and how to manipulate and deceive. These are skills that young animals have to master if they are to have any chance of advancing in the game of survival.

SYNCHRONIZED SOCIALIZING
A coalition of Atlantic spotted dolphins in the Bahamas practising synchronized swimming. They are highly social mammals, with strong friendships and family ties and a culture that includes active teaching of the young.

FOOD FIGHT
Previous page Most animals try to avoid high-risk fights, especially if dangerous weapons such as talons are in use. Here a bald eagle dives on another smaller, lighter one that has pulled a salmon carcass from the Chilkat River, Alaska. The smaller eagle knows it is not worth risking injury to fight back and so gives up ownership.

Power

For most animals, aggressive interactions are part of life, but to scrap all the time wastes valuable energy and risks injury or even death. So animals have ways of avoiding direct confrontation or repeated challenges. In groups, members are either dominant or submissive relative to each other and so can avoid the need for repeated high-risk confrontations. This order is maintained by threats and displays and only changes when a subordinate successfully challenges a dominant animal.

A high-status individual reaps a disproportionate benefit from its position and can ultimately expect to produce many more offspring than subordinates. For example, during times of water shortage the highest-ranking female vervet monkeys have greater access than subordinate females to water in tree holes. High-ranking chacma baboons have the first access to prey that has been caught by the group. And high-ranking male bonnet macaques partake in most of the mating.

For subordinates, sacrifices can be extreme. As well as being the last to feed or drink and suffering the constant low-level stress of life in the ranks, they may never have a chance to breed. But if a subordinate is closely related to the dominant individuals, assisting them may have a genetic benefit, as some of his or her genes are still passed along in the offspring. Also, by staying out of trouble and slowly moving up the rankings, they may outlive those above them, though it might be a very long time before a top spot opens up.

But subordinates can use their time in the lower ranks to acquire skills, resources and attributes that will help their prospects. Wild dogs, for example, are always practising hunting skills, and they may even learn how to be a parent by looking after siblings. Meerkats form political alliances, and savannah baboons display a number of mating tactics correlated with their age, for example, forming alliances to combat higher-ranking males and have a chance of accessing females for copulation. Subordinates can also try to subvert the system by adopting 'sneaky' strategies, especially in a large group with several males, where it may be hard for the highest-ranking one to control all the mating opportunities. Weaker males sometimes pretend to be females not just to avoid confrontations but to gain access to real females right under the nose of the dominant male. But for the most part, behaving submissively is only any good as a short-term strategy.

For a young animal wanting to increase its status, the decision to initiate a fight is generally made when a higher-ranked animal is no longer viewed as unbeatable, perhaps because of injury, illness or age. Then the behavioural interactions that have kept the peace break down, the challenger responds aggressively to the dominant's ritualized displays, and the confrontation escalates into a battle that may result in regime change.

But success has its problems. An individual must actively defend its place as well as continuing to challenge those even higher up. At the very top there's a lot to lose, and defending territory, corralling mates and guarding other resources can cause high-ranking individuals to lose condition. Animals in the top spots have high levels of the stress hormone cortisol, which can potentially knock years off an animal's life.

In the end, accumulating power comes down to balancing the costs of a struggle and the benefits of pursuing different strategies. And, of course, those with the power have the greatest chance of leaving a genetic legacy.

FOX FIGHT
An Arctic fox asserts his dominance over a rival male, possibly his brother, as the courtship season begins, and mating becomes as important as foraging for food.

Social life can be doubly testing for young eagles

Bigger, older birds let starving young ones do their work for them

WHY BIG BIRDS WIN
An adult bald eagle diving in for the attack. The heavier more experienced adult will always win the salmon prize in any skirmish with younger, lighter birds.

BULLY TACTICS
A flying tackle from an adult. Physical contact is rare among the eagles, but this adult is making sure that it gets the salmon that the youngster has just dragged out of the river.

In Alaska, the odds are stacked against young bald eagles. Surviving the bitter cold of their first winter comes down to finding enough food, and since the eagles feed mainly on fish, this becomes increasingly difficult as rivers and lakes freeze over. Starvation is a real danger because young birds simply do not have enough time to develop hunting skills before they are confronted by the full force of an Alaskan winter.

Their best chance is to find somewhere with an accessible food supply. One such place is an 8km (5-mile) stretch of the Chilkat River in southern Alaska, which remains ice-free even as temperatures plummet to -30°C (-22°F) and below. The cause is a natural reservoir – a fan-shaped accumulation of gravel, sand and glacial debris – that fills with melt water in summer. As winter bites, the reservoir water remains 6–11°C (10–20°F) above surrounding water temperatures, and as it slowly percolates into the Chilkat River, it keeps this stretch from freezing.

This quirk of nature offers the bald eagles a winter lifeline. Salmon use the short stretch of open water as a spawning channel through the autumn, but it's the late run of large chum salmon that is the magnet for eagles, with fish sometimes arriving in large numbers well into December. Bordered by mountains and with heavily forested slopes and riverbanks, the valley also provides roost sites and shelter from winter storms. Even when gales are blowing at the mouth of the river, it is calm upstream.

But so desirable is the valley that it can get very crowded as eagles of all ages converge here from across southern Alaska. Most winters see at least 2,000 eagles, and in some years as many as 3,500 are drawn to the valley – possibly the largest assemblage of eagles anywhere. So the competition the novices are up against is intense.

At its pre-Christmas peak, the Chilkat's riverside trees are filled with eagles of all ages. But very soon the temperatures plummet to -30°C (-22°F). The eagles can't afford to waste energy, and so they remain perched for hours, not moving, expending the minimum amount of energy.

In the river below, hundreds of dead and dying fish are crowding into the open stretches of frigid water. A single chum makes a substantial meal – large ones weigh

in at 10kg (22 pounds) and are up to a metre in length. But to catch one in the deeper water is not possible, and to drag one out risks getting wet. Also, once a chub is grabbed, an eagle seldom has it to itself.

The younger birds are usually the hungriest and the most willing to take risks. A desperate eagle will eventually swoop down to the water's edge to drag out one of the rotting carcasses or to wade into a fast-flowing stream to grab a dying fish with its powerful talons. It can be quite a battle to get one of these huge fish out of the water, and keeping hold of it is even trickier.

A whole fish is also too heavy to fly off with and so must be partially consumed on the riverbank. A young eagle with a fish is a sitting target for the bigger, older, more aggressive birds. It is almost always forced to stand aside and wait until the challenger has eaten its fill, in the hope of retrieving scraps or a rotting head or tail.

With the adults, it's hard to predict the winner of a confrontation.

WINTER REFUGE
The confluence of the Chilkat River (left to right) with the Tsirku (centre), Alaska, where several thousand eagles gather in autumn and winter to take advantage of the open water and the salmon gathered in the shallows after spawning. Warmer water seeping in from a meltwater reservoir keeps this stretch of river from freezing.

FROZEN HANG-OUT
Eagles roosting in the cottonwood trees along the Chilkat River. They fly down to try to haul out a salmon only when hunger overcomes the need to conserve energy. The young birds will be the ones most desperate to try their luck.

DAWN WARM-UP
Two eagles survey the Chilkat River. By midwinter, when the salmon run has finished and most carcasses are inaccessible, just a few resident eagles stay put.

CATCH OF THE DAY
An adult eagle drags a salmon up onto snow-free gravel. It will toss back its head and vocalize to warn off any other eagles bold enough to try to take its find.

THE BIGGEST GET THE BEST
Overleaf The young eagle (left) has pulled out the salmon, but the biggest adult has grabbed the prize, and the others will take any leftovers.

When challenged, some eagles simply abandon their kills while others beat their opponents with their wings and lunge at them with bared talons. The most spectacular confrontations occur when one eagle swoops in and catches a feeding eagle unaware – crashing into it and sending it somersaulting away into the snow. Large adults are usually the most successful, but feeding eagles are more likely to hold onto their fish if they see the challenge coming and if they display by calling while in possession of the carcass.

The salmon run peters out after Christmas, and by February, a shelf of ice has built up from the edges of the open channels to a point where the remaining salmon carcasses are either frozen in the ice or are in water so deep they cannot be reached. The adults that remain in the area are, thanks to the salmon, in prime physical condition at the onset of the breeding season.

The immature eagles are not so lucky. They are forced to disperse away from the river, usually heading southward into British Columbia and even hundreds of miles away into Washington State in search of food. But the bounty of the Chilkat at least gives them a chance of getting through their first few winters, though it will be several before they are ready to breed.

Growing up by showing off

If a young grouse is ever going to succeed, he has to learn to perform on the dance floor

There comes a time in the life of a young male sharp-tailed grouse when he has to take a huge leap forward in the survival game and start to compete with the grown-ups. If he is to have any chance of contributing to future generations, then he must enter in the race to win a mate. And that carries with it a chance of failure and also a risk of serious injury.

The time for these young challengers to test their powers is spring, when on the sage-covered prairies of eastern Montana snow still lies on the ground. In April, the lengthening days trigger something in the brains of the males that changes them into strutting, dancing show-offs.

Their overriding desire is to win a mate – or several. And the way the males choose to impress the ladies is to gather together in a dance competition, where the females are the judges. Such gatherings are known as leks – adopted by many animals, from insects to antelopes. But it's among the birds that this style of mate choice is best known, particularly among game birds.

A dance arena for the sharp-tailed grouse covers a tennis-court sized area of dry, open ground. These are traditional sites, often used year after year, and the trampled vegetation makes them easily recognizable among the sagebrush. As the winter snow begins to melt, more than 20 males may gather at one of these arenas, and in the pre-dawn light, begin to dance and squabble over ownership of the prime central spots.

The dance routine involves a male bending forward with head and wings held out horizontally and short tail held erect, the central feathers bringing it to a point (hence the name, sharp-tailed grouse). He also sports bright yellow combs over each eye and purple sacs on each side of his neck, all of which are inflated during the display – designed to both intimidate rivals and attract the ladies.

If one male intrudes into a neighbour's territory or challenges him, then the two rivals rush forward or circle while rapidly stamping their feet – about 20 times per second – clicking their erect central tail feathers, and emitting hooting, clucking, cackling and gobbling sounds.

SHARP-TAILED BATTLE
Two male sharp-tailed grouse jockey for position on the lek – the traditional grass display arena. Fights can be prolonged and vicious, the cocks clawing at each other's eyes and throats, leaping and tumbling as they try to displace rivals.

A female takes her time over her choice and may visit over several mornings, eyeing up the males as they dance and fight

The confrontation may end in a restrained face-off, as they stare at each other until one backs down. But if they are more evenly matched, then a full-blooded battle may develop.

The combatants beat each other with their powerful wings, claw at each other and dart in at lightning speed to peck at eyes and throats. These fights can be prolonged and leave feathers flying and bloodied birds chasing back and forth across the lekking ground. This is the competitive cauldron into which the young birds must step. Only by entering the arena will they have a chance of winning a mate. Yet only the dominant males end up holding the central positions in the lek and attracting the females' attention.

These dominant males tend to be significantly larger and heavier than the younger birds and so are likely to have more stamina and stay longer on the lek, display more vigorously and fight harder. The less experienced subordinate males end up on the periphery of the lek, where they will have little chance of catching the discerning eyes of the hens.

Female grouse begin visiting the lek once the males have established themselves in a hierarchy. A female takes her time over her choice and may visit the lek over several mornings, eyeing up the males as they dance and fight. If she steps into the arena, she is harassed and displayed at by the males – often en masse and with such vigour that she is forcibly driven away. But this is all part of the game.

Her choice will almost inevitably be one of the males who holds a central territory. For female grouse, the benefits of having all the males performing in a confined space means they can compare potential mates relatively quickly and without having to travel far – which may also reduce the time she is exposed to predators. Once mated, she has no more to do with her choice of male and leaves to nest and raise the young. The males, meanwhile, continue with their strutting and scrapping in the hope of impressing other visiting hens.

The younger birds out on the periphery of the lek have to battle all comers as they try to manoeuvre themselves towards the centre of the lek. It may take several seasons before they can win one of the favoured spots. Though they do have the remote chance of intercepting a female as she moves through the lek – *if* they can avoid being spotted by a dominant bird and beaten up. On the outskirts, there is also a greater chance of being taken by a predator. So it's tough being young. But for the females, such competition and exhibition of male skills means that they have a chance of selecting the strongest and fittest males to father their offspring.

THE FEMALE WINS THE PRIZE
Top left A younger male attempting to displace an older, larger male with a full-on attack.
Top right The combatants clawing at each other with their talons. Feathers will fly and blood may be drawn.
Bottom left Rival males displaying at the edge of the lek, sharp tails straight up and rattling, pink-purple booming sacs showing as they vocalize, and bright yellow eye-combs prominent.
Bottom right A female standing at the centre of the lek watching the males displaying. It may take her several visits before she selects a male, usually one holding a central territory.

Archery lessons

Watching your elders gives you skill, and youth gives you the speed to snatch their prey

Lots of fish eat insects. Trout will snatch bugs from the water surface, a habit exploited by fly-fishermen. The Amazonian arowana will occasionally jump clear of the water to grab insects from riverside vegetation. But archerfish have taken insect-catching to a whole different level. Living in the brackish waters of Asia's and Australasia's mangrove swamps, archerfish hunt bugs, spiders and even small lizards by shooting a jet of water from their mouths with pinpoint accuracy, knocking their victims from streamside perches, then gobbling them up when they hit the water.

This unusual hunting strategy was first recorded as long ago as 1764, but it's only in the past few years that its true complexities have been revealed. At first glance, these fish don't look much like deadly predators, but their anatomy offers some clues to their specialist hunting skills. For a start, their dull, mottled markings and very narrow bodies make them hard to see when viewed from above as they lurk in the

MASTERCLASS
A master archer preparing to strike, watched closely by youngsters. They are learning not only technique but also where the prey will fall, aiming to dart to the surface and grab it before their teacher can.

FIRE-POWER
The master shooting a powerful jet that may reach several metres (more than six feet), though is most accurate at closer range. A master knows the exact angle and intensity of jet needed for the size of the prey – a judgement that only comes with practice.

sun-dappled waters beneath mangrove vegetation. More unusually, their backs are almost straight from the tip of the snout to the dorsal fin, which means they can swim right up to the water's surface without giving away their presence.

Their eyes, too, are special. Not only are they huge but they also contain specialized receptor cells in the retina that greatly increase their ability to distinguish different colours: the brown of an insect against a green background, for instance. And the relative position of their eyes means they have a form of binocular vision – very rare in fish – that helps them to judge distances accurately. But it's not their physical specializations but rather their mental abilities that set these fish apart.

The first problem they must solve when hunting is how to spot suitable prey and then take aim through the distortions and ripples of the water surface. This is complicated because light bends at the interface between water and air, an effect known as refraction – an optical illusion that appears to shift the position of an object on one side of the boundary relative to the other. Experienced archerfish are able to compensate for this, locking onto victims, irrespective of their position and angle relative to the prey.

Once an archerfish has locked onto its target, the next challenge is to create a narrow jet of water and shoot it with pinpoint accuracy. It does this by first raising its tongue against a groove in the roof of its mouth to form a long, narrow tube. It then squirts a mouthful of water along this tube by rapidly snapping closed its gill covers. This water pistol can generate enough force to fire a narrow jet of water up to 3.5 metres (12 feet), though this jet is most accurate when the shot is fired from less than 1.2 metres (4 feet) away.

And there's more to this operation than simply hitting the chosen target. The archerfish must also assess the size, type (and 'grippiness') of any prey item – a praying mantis is a very different proposition from an ant or a butterfly. The water jet must be perfectly judged – sufficiently strong to knock the insect from its perch and into the water but not so powerful that it catapults the victim off into the surrounding vegetation or towards any lurking competitor. Like all aspects of the archerfish's hunting technique, this fine judgement only comes with

practice, and inexperienced fish must often make spit after spit before finally dislodging the prey.

Even after a successful strike an archerfish's difficulties aren't over. In these mangroves, there are lots of opportunists on the lookout for a free meal. So an archerfish must also be extremely adept at retrieving its prey – and that's where its predictive skills and lightning-fast reactions really come into play.

It's been shown under laboratory conditions that an archerfish reacts to the motion of falling prey and computes the spot where it will fall within 40 milliseconds of the initial hit – so quick that it's almost impossible for competitors to outpace the shooter, unless of course that competitor happens to be another archerfish. Which is exactly the situation young archerfish face.

For safety reasons, youngsters often hang around in small shoals. So after a successful hit, there's inevitably a frantic scramble as fish race to be first to reach the tumbling insect. Competition is so intense that the successful shooter has a 50 per cent chance of losing its meal (and the bigger the shoal, the greater its chance of going hungry). But perhaps the most remarkable aspect of all this is that archerfish aren't born with all these talents ready to go. Though they can spit from an early age, they are initially wildly inaccurate when they start, unable to compensate for the refraction of the water or to judge the correct force required to dislodge different types of prey.

The subtleties of hunting must be learnt, and the best way for a young fish to improve its technique is not just to practice but to find itself a good teacher – an older fish that has perfected the art of spitting. Young fish that have the opportunity to watch a 'master archer' in action quickly improve their own techniques and can shoot with far greater accuracy and more successfully than a young fish raised in isolation.

SHOOT, HIT AND SNATCH
Left to right **A master archer eyeing up a butterfly. It almost instantly programmes the angle and the intensity of the jet needed for the size of prey. As the butterfly is knocked off its perch, the archer darts forward to grab it before one of the watching youngsters gets there first.**

Youth among the superbeasts

Where Africa's biggest buffalo meet its biggest lions.

FEARED AND FEARFUL
A large male African buffalo – a formidable opponent for any hunter. Both males and females use their horns as weapons, but those of a mature bull are fused at the base into a shield of bone across the forehead.

GAME ON
A group of lionesses openly approaches a large herd of bulls, aiming to pick out a vulnerable individual. Lions and buffalo live cheek by jowl on this Okavango Delta island. Deprived of normal prey, the lions have become buffalo-hunting specialists.

A few years ago, a shifting river course created an island in the Duba Plains, deep in Botswana's Okavango Delta, trapping more than a thousand African buffalo on just 200km^2 (78 square miles) of marshy grassland. At the core of any buffalo herd are groups of related females and their offspring, and the same is true here, except that the herd is much larger, and mixed in with it is another group – the bulls.

A third element of the buffalo population are the young bulls, unable to migrate and with challenges and threats from all sides. They live on the fringes of a society ruled by strict hierarchies and where life is a constant battle for enhanced status.

Only the biggest bulls win a harem of females, and these bulls are intimidating. Standing 1.7 metres (5 feet 7 inches) tall at the shoulder and weighing more than 900kg (1,984 pounds), they are capable of reaching speeds of 55 kilometres an hour (34mph), at least over short distances. Add some fearsome weaponry – razor-sharp hooves and great hooked horns that can measure more than a metre from tip to tip – and you have one of the most feared animals in Africa.

It can take several years before a subordinate bull can even think about going up against one of these monsters. In the meantime, he joins one of the bachelor groups that track the main herd. Here he will hone his combat skills, trying to advance through the ranks and readying himself to challenge for one of the top spots.

Most of these adolescent confrontations are settled quickly with threats and bluster, but if two young bulls are evenly matched, contests can escalate into protracted and violent confrontations. The bulls will charge at each other, lowering their heads to take the impact on their massive horns and heads. Pushing and shoving, hooking and goring, the battle is likely to end with the weaker contestant being knocked to the ground in a cloud of dust.

But on the Duba Plains there is another element in the survival equation. As the waters rose to create the island, several prides of lions were also trapped, and a uniquely intense struggle has developed between hunters and hunted. Deprived of their more normal prey – zebra, giraffe and impala – the Duba lions have developed

strategies to tackle the aggressive and often unpredictable buffalo.

Lions normally hunt in the cool of the night, but on Duba the lions often stalk their prey in the midday sun. They have also adopted a most un-lion-like behaviour – wading and even swimming through the thick papyrus swamps and deeper floodwaters to track the buffalo around the marshes. And to tackle such formidable prey, the Duba prides have become pumped-up supercats that now dwarf other lions. The lionesses, for example, are almost the same size as male lions elsewhere in Botswana.

As for the buffalo, they not only live in a huge herd, they also work together to defend themselves. When they need to rest, as they frequently must to chew the cud, the larger animals will create a protective cordon, gathering together and lying facing outwards to form a defensive wall.

But the buffalo can't be vigilant all the time. Either by ambush or by direct charge or through a war of attrition, the lionesses will usually find some way of singling out a vulnerable member of the herd. Even if they can drag down their chosen prey, the hunt is far from over.

Despite the rivalries between bulls, when under threat they tend to stick together. The younger bulls

SUPERCAT
A lead lioness crashes through water in pursuit of a buffalo. Hunting for long periods in Duba's marshy environment has given these lions heavily muscled upper bodies. Indeed, they are giants among big cats, larger than any others found in Africa.

▲ **LION TACKLE**
A lioness getting to grips with a female buffalo. The hunting party is invariably a group of females working together to ambush, chase and pull down their victim, and it usually takes several lions to bring down a full-grown buffalo.

▶ **STRIKING BACK**
Bachelors ganging up on a lioness. Despite their rivalries, young bulls stick together to see off lions and are capable of killing isolated adults.

often take the lead, and it's common to see several of them charging again and again into the cats, trying to drive them off and free a downed buffalo, allowing it to regain the safety of the herd.

These aggressive counter-attacks have been recorded in more than three-quarters of all the lion hunts observed on Duba. Even more surprising, the buffalo will launch pre-emptive strikes on the lions, especially if they stumble upon one isolated from the pride or find young cubs. Often attacking in strength, they will attempt to hook and trample the unfortunate lion – dispatching it on the spot or consigning it to a lingering death through infection.

In such an intense arena, young bulls have to be constantly vigilant. They can't afford to drop their guard for a moment, both because of their ongoing battle for self-advancement within the herd hierarchy and the constant threat from some of Africa's most formidable lions.

UNEASY STAND-OFF
Cows and bulls looking on as a pride feasts on a member of their herd. In more than three-quarters of the lion hunts, the buffalo will counter the lions' aggression, sometimes fatally injuring their hunters. Any cubs they encounter will be trampled to death. Though about 20 buffalo are killed a month, the population is thriving, its destiny tied in with that of the lions.

When chimps' tools become weapons

From winkling out grubs with sticks to spearing bushbabies

There are many examples of non-human animals using tools. The Galapagos woodpecker finch uses cactus spines to extract insect grubs from dead wood and may even modify the spine. Dolphins carry sponges in their mouths to protect their beaks as they sift through sediment on the seabed looking for fish to eat. Sea otters will strike abalone shells onto anvil rocks balanced on their chests as they lie in the water. There's even a desert-living ant whose workers pick up stones in their mandibles and drop them down the entrances of nests of rival colonies, stopping their workers from getting out to feed. But it's among the primates and especially the chimpanzees that the most advanced tool-users are found. Chimps have been seen to crack hard nuts using rocks as hammers and anvils, to soak up

STONE-THROWER
A male standing up, hair on end, about to hurl a rock at a rival. The Fongoli chimpanzees will frequently stand up to get a better look and will even walk bipedally occasionally when it makes travel easier.

TERMITE-FISHER
Opposite page A young chimp fishing for termites, using a tool fashioned from a twig to extract the grasping insects from deep within their earthen mound. It has learnt from its mother how to extract these protein snacks.

water using shredded leaves and to fish for termites and ants with specially collected and prepared sticks and grass stems. In Senegal, West Africa, one population – the Fongoli chimps – has developed an even more surprising tool-use culture, and it's the younger, subordinate animals and the females that have, by necessity, led the way.

The Fongoli chimps inhabit a savannah dotted with patches of grassland, rocky plateau and sparse woodland. The environment is brutally hot during the dry season and battered by torrential storms in the wet season, which has forced the chimpanzees to behave very differently from their rainforest relatives.

They range farther than any chimpanzees so far studied – over more than 60km^2 (23 square miles). Led by the older males, they spend a lot of time on the ground, often walking ten kilometres (six miles) or more in a day, sometimes standing upright to better scan their surroundings. As surface water evaporates in the intense heat of the dry season, the older animals know where to dig wells in the dry riverbeds, using their hands to scrape away the dirt and rocks and then scooping out the water. Subordinate chimps wait their turn to drink. The Fongoli chimps have also taken to spending time in caves, where they socialize and snooze in the heat of the day.

Food is always in short supply, and compared to forest chimps, they spend a lot of time fishing for termites – a valuable source of protein – eating them year round. But in the dry season, other innovations are required. The dominant male chimps hunt green monkeys and even baboons, but compared to other chimpanzee societies, they seldom share the meat, possibly because it is much harder to obtain. So the physically weaker females and subordinates have been forced to find another source of protein: bushbabies. These small, nocturnal primates hide in hollows in trees in the day, but the Fongoli chimps have come up with an innovative way of catching them – using spears.

When a chimp finds a potential bushbaby hiding place, usually a hollow tree limb, it begins to search for a suitably sized branch – about the diameter of a broom handle – and then breaks off a length, usually about 60cm (23 inches) and strips it of twigs and bark. The spear is then thrust repeatedly into the hollow where a bushbaby might be lurking. The hunter examines the tip repeatedly between stabs to see if it has hit prey, and then, if it has, the chimp breaks into the hollow and drags out the bushbaby to be eaten.

At least ten female and subordinate chimps in the Fongoli population have been seen hunting like this, and others have been seen fashioning spears. Young chimps will watch older ones at work, and one young male was even seen using the spear playfully, possibly learning the technique. It's the first record of any non-human mammal killing another using a fashioned weapon – a testament to the mental power of a fellow primate.

SPEARING A BUSHBABY
Top left Stripping leaves off a suitable length of spear stick. Watching the female at work is a youngster.
Top right Stabbing repeatedly down into the hollow branch.
Bottom left Extracting the live bushbaby prey.
Bottom right Holding the bushbaby before quickly killing and eating it.

COOLING OFF
Overleaf A group of adult chimpanzees soaking in a waterhole in the rainy season. Life for these savannah-woodland chimps can be very hot, and lying in water is one way to cool off. It can also turn into a social event.

First cooperate, then assassinate

How to become queen of the honey ants

HONEY-LARDER HANG-OUT
The honeypot workers hanging deep in the nest. These are the largest workers, which are force-fed nectar, protein and even water until they swell to the size of cherries. Hanging up in storage, they can feed the colony when lean times arrive.

QUEEN OF ALL THE COLONY
A honey ant queen – now mainly an egg-layer – surrounded by workers tending to the larvae and pupae. Above her hang living-larders. She may live for 20 years or more, supplying the colony with workers. The queen's chamber and the nurseries are deep underground, to escape the desert heat.

Success doesn't always come from being the best fighter or hunter. Sometimes it's the exploitation of others that gets you there. Queen ants are normally very aggressive, but when a future melliger honey ant queen sets out on her life journey, her behaviour can be the very model of cooperation. That journey begins, in the southern United States or Mexico, with the eruption at the end of the day of thousands of virgin queens and princes from nests all across a desert environment, triggered by the brief rains. It's a mass aerial emergence that ends with the females mated and the males dead. By daybreak, the mated queens have dropped back to earth to found new colonies.

The rain that triggered the emergence also softened the soil just enough for the queens to begin digging. One will start and will soon be joined by another. The imperative of digging a chamber before the sun bakes the soil rock-hard means that cooperation is the only way to succeed, and several queens may form an excavation team as the digging becomes a race against time. This cooperation is repeated all across the desert floor as thousands of new nests are dug overnight.

Once sealed inside their humid subterranean cavities, each young queen now lays her first batch of tiny eggs. Her only sustenance for the next few weeks will be her stored fat reserves and metabolized flight muscles. When the first tiny workers emerge from their cocoons, some tend to the egg-laying queens, and the others go to the surface to find food for the starving queens and the growing colony. But the desert floor is littered with new honey ant colonies, and there's not enough room for all of them. So the workers are forced to eliminate neighbouring nests. Only the biggest colonies, those with the greatest number of workers – in turn determined by the number of founding queens – will survive the carnage.

But while the workers are doing the dirty work above ground, a more insidious fight for domination is beginning under ground as the formerly friendly royals begin the battle for top spot. It's an unequal battle. The weaker queens appear to crouch submissively before the more dominant. The workers start singling out the submissive ones, first harassing them, then attacking and tearing them to pieces. Nothing goes to

waste in the colony, even a royal carcass, which is fed to the developing larvae, some of which may even be the dead queen's own offspring.

Only the most dominant royal seems immune to attack by the workers, and at the end of the purge, she will be the only one left, and all the workers will be loyal to her. She may live for 20 years in her underground bunker, ruling over tens of thousands of workers in a colony that may grow to dominate a huge desert empire.

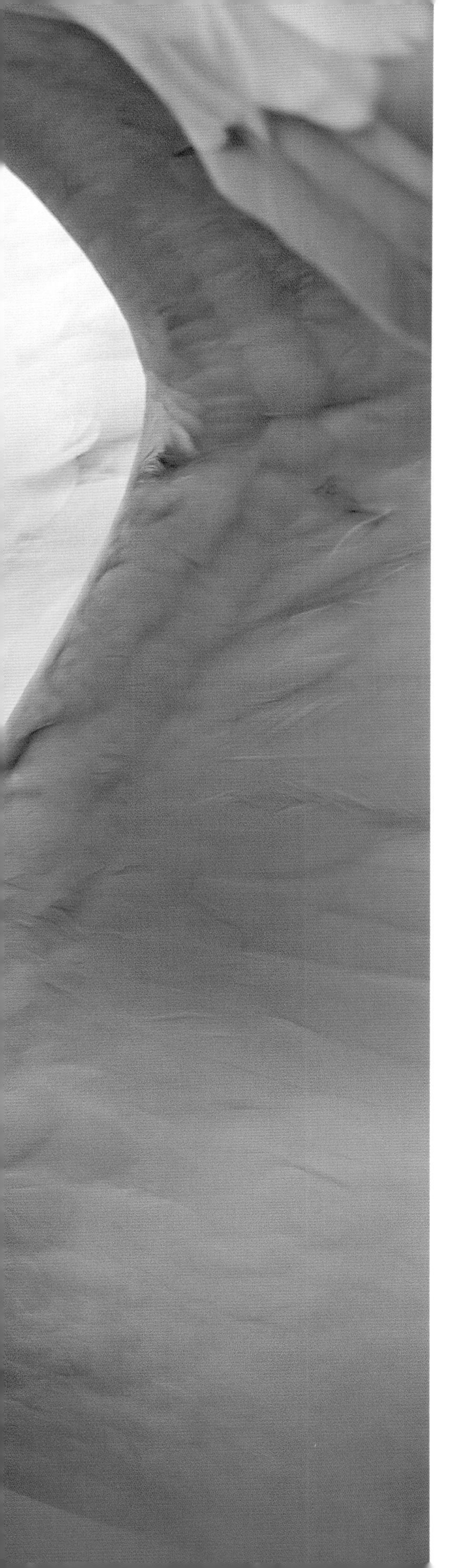

5 Courtship

Courtships and partnerships in the animal world rarely conform to our romantic ideals. Though the natural way to win a mate may involve spectacular and beautiful displays, the giving of gifts and even passionate embraces, it can also include deception and infidelity and sometimes violence.

BONDING MOMENT
A silverback mountain gorilla embracing one of his females following mating. He will fiercely guard his harem against male rivals but is gentle with his family.

FLORAL OFFERING
Previous page A male gannet offering red campion to his mate as a courtship gift.

Courtship

To stay with the same partner is rare. Only in birds, and then only a handful of bird families, including albatrosses, swans and geese, do the same partners meet up year after year. The wandering albatross is probably the most impressive of those, pairing for life – a relationship that can span well over 40 years. More than 90 per cent of birds show apparent monogamy – staying together for at least a season to rear the family. That they do this is in part because the father is equally capable of feeding the young (unlike mammals, where only mothers can produce milk). But there have long been clues that apparent monogamy doesn't mean fidelity.

Snow geese come in two colour phases, white or blue, and the regularity with which pairs of pure white geese produce some blue goslings indicates that all is not as it appears. DNA fingerprinting reveals how common it is for birds to cheat on their partners – from sparrows to even some species of albatross. So in this battle of the sexes, the female may choose to have it both ways, pairing with a good father who feeds her chicks and mating with other males to obtain different or better genes – indulging in a little genetic pick and mix.

Across all animal groups it is usually the male who puts in the most effort to impress his mate and the female who is most discriminating in her choice. This is because it's normally the mother who invests the most in her eggs or young, while males can and do get away with simply inseminating females and taking no further part in raising the family.

The most extreme examples of this approach are birds of paradise and bowerbirds, probably because in the tropical rainforest where they live there is plenty of food to allow male extravagance and to support single mothers. Males spend weeks and months displaying for intensely choosy females, who after mating have nothing more to do with them.

But in some bird species and a few other animals where males show paternal care, a male may also have to prove his value as a father. Terns present females with gifts of fish, and male European wrens and weaver birds build nests to demonstrate their DIY skills to prospective mates.

Among mammals, the females often let the males prove themselves by encouraging outright male v male competition, in which the combatants may even put their lives on the line to win the prize of fatherhood.

Blackbuck and fur seals often have to fight viciously to hold a territory, and the females seem to select a male on the basis of the position that he holds rather than his looks. Males may literally die for love from their injuries or exhaustion or, in their weakened state, be picked off by predators. But the rewards for the last warrior standing can be high. A winning male fur seal may mate with 10 or 20 females in a season, while his rivals get none.

If you are too small to take on the big guys directly then deceit can be the best strategy. Smaller male fish and garter snakes will even mimic females to get close to a courting pair and inseminate the females under the very noses of the bigger males.

But it's among the spiders that the males can be under the greatest pressure to perform. One wrong move by the beautifully coloured peacock jumping spider as he performs his wonderfully intricate dance and he switches from suitor to supper.

A GREATER PARADISE DISPLAY
A male greater bird of paradise in the final stages of his courtship display. Males group together to perform, and a female selects her partner on the basis of both his looks and his performance. After mating she has no further contact with him.

The male weevils that win by a neck

When boring is sexy

Size often matters in the mating game, and at first sight it would seem to be the overriding factor in the breeding success of one bizarre-looking insect – the giraffe weevil. The male's elongated rostrum or 'nose' projects beyond its eyes, accounting for 50 per cent of the length of its body (giving it its common name), and is topped off by a pair of antennae and pincer-like mouthparts. The largest adult males are six times the size of the smallest, but despite this differential, it's not always the goliaths of the weevil world that succeed in fathering offspring.

The giraffe weevil is in fact a type of beetle, the largest in New Zealand, and frequents the trunks of dying trees. Both sexes have considerable size variation, but the males tend to be significantly larger overall – up to 120mm (4.7 inches) in length, compared with the females' maximum length of 50mm (2 inches). The insects spend two years as larvae, boring through wood, and eventually emerge in their second spring. As adults they live only for a few weeks, camouflaged against the brown bark, but they make the most of their adult season, gathering together in breeding groups of up to a hundred.

The female differs from the male in that her antennae are farther down the rostrum, halfway between its tip and her eyes. This probably keeps them out of her way while she uses the end of her rostrum to bore into the wood, excavating a hole for her eggs. As she digs, she moves her head rhythmically from side to side to slice through the wood, stopping every so often to eject the accumulated debris from her rostrum and using her antennae to delicately flick it away. Only when the hole is 3–4mm (0.1–0.2 inches) deep does she turn around, lay her eggs in it and pack the entrance with the excavated material. The whole process takes about half an hour.

It's when the female drills into the bark that a male has to try to mate with her. This means that the sexiest thing a female can do is to start boring, and the first passing male of any size will be compelled to stop and stand guard over her, so he can fertilize her eggs before she lays them. Of course, the pair don't remain alone for long. An aggressive audience of males of various sizes soon assembles – and this is where the giraffe-like necks come in.

JOUSTING RIVALS
Drawn by the scent of a female, rival weevils are gathering to wrestle for her favours. In the world of weevils, size matters, and it pays to have a very long 'nose', or rostrum. At the end of the rostrum are pincers that are deployed as weapons.

A larger male looms up and uses his long rostrum to rake across the guarding male's body, trying to wound or dislodge him. If this doesn't work, he uses his mandibles to bite at his opponent's limbs. The male with the longer rostrum can deal more powerful blows while keeping his rival 'at arm's length' and reducing any risk of injury to himself. The fight ends when one of the males withdraws or is thrown from the tree.

In one-to-one encounters, small males can't compete with large ones and avoid direct confrontations. So how can they succeed in mating? What they can do is watch the battle of the giants and bide their time. If the two contenders become distracted and leave the female unattended while they fight, the small male can dash in and mate with her.

Even when a large male is guarding a female and is already fertilizing some of her eggs, all is not lost. The small male can flatten his body and lower his rostrum to appear less male-like and slip in, unnoticed, under the legs of the giant. He will then gently mate with the female while the large male stands blissfully unaware over both of them. So at least some of the eggs she is about to lay will be fertilized by the small but sneaky male.

THE GUARD AND HIS BORER
A female is boring a hole in wood with her rostrum. Standing guard over her is a large male, aiming to ensure that only his sperm will fertilize her eggs. When she has finished boring, she will turn around and lay her eggs in the hole.

Heavy sex

Why a female turtle might wish she wasn't so attractive

A male green turtle's approach to courtship is so competitive and intense that he subjects the object of his passion to an ordeal that seems to put her in imminent danger of drowning.

A suitor's first move is to intercept a female, and the time and place to do this is when she is ready to lay her eggs and is approaching the nesting beach (normally where she hatched, possibly 40 or 50 years ago). When a testosterone-fuelled male finds an unattached female, he briefly courts her by approaching head-on and nosing and nipping at her face. If this results in her acquiescence, she will allow him to swim around behind her and climb aboard, using the claws on his flippers to attach himself on top of her shell, his concave underside fitting snugly over her convex upper shell.

He has a long tail with a horny claw at the tip, which he hooks under the female so that he can transfer his sperm via her cloaca – the opening at the base of her tail through which she will eventually lay her eggs. If she wishes to deny him access to her cloaca, she can press her hind flippers together or cling to the seabed while pushing her rear end into the sand. And bearing in mind what can happen during the mating season, it is not surprising that females sometimes resort to doing this.

Once the pair are securely attached, they swim through the warm tropical waters, locked together in their embrace. But the honeymoon doesn't last long. If there is a male nearby who has not been fortunate enough to have found a mate, he will be intent on breaking up the couple.

He will manoeuvre himself into position at the rear and with his serrated beak will try to take chunks out of the first male's tail and flippers before switching his attention to the more delicate skin at the base of the tail. The male riding on the female's back will take his punishment and hang on grimly. But things can get worse.

Other males may arrive and join in the attack. Green turtles, being reptiles, have to breathe air, and by now the female is beginning to become exhausted, trying to escape her pursuers while doing the swimming for two.

A BURDEN OF MALES
A male green turtle clasping another male, trying to dislodge him from his mating embrace with a female. She now has to swim carrying a double load, struggling to the surface to breathe every few minutes. If the rival dislodges her first suitor, there is a chance that some of her eggs will be fertilized by the second male.

She needs to get to the surface every few minutes, and this becomes more difficult as more suitors join in the chase. Sometimes a male will become so desperate that he will attach himself to the first male, and then another one will do the same, and so on until the female has up to four passengers.

Getting to the surface to breathe with the weight of four 180kg (400-pound) turtles on your back is a desperate struggle. These orgies can last up to 12 hours before the female is eventually released and can swim to shore to lay her now-fertilized eggs.

A word of warning: it's not just female turtles that may have to cope with the males' weighty and horny embraces. When pumped up in the breeding season, males are so indiscriminate that they can be caught by fishermen trailing a wooden decoy attached to a line and then reeling in the male turtle attached to his newfound love. And there are even cases of lovesick males attempting to attach themselves to human divers. As human swimmers don't have the breath-holding abilities of turtles, they would be well advised to keep an eye out for the lovelorn and distinctly 'hot-blooded' reptiles.

THE CHASE
An embracing couple is being chased by half a dozen or more amorous suitors, all trying to dislodge the male. The pursuit can last for hours, making it harder and harder for the female to surface to breathe.

Flashy fiddlers risk all to take a female home

What a female crab wants is a place to live and lay

WELCOME TO MY BURROW An inconspicuously coloured female fiddler crab standing over the entrance to a burrow occupied by a male, whose claw is extended to tussle with a rival. The other male is challenging him for the ownership of both the burrow and the female. A jousting match will follow.

THE COME-ON GO-AWAY WAVE Flashing his giant white claw behind his bright blue-black back, a male fiddler crab is sending out a powerful visual signal. To males it means keep away, but to females it's a come-hither wave.

For some males, it might not be enough to be the biggest or the most colourful to attract and keep a mate. You might have to provide a new home and wrestle with competitors – and risk your life in the process. That's certainly the case with the male elegant fiddler crab.

This little fiddler inhabits the highest reaches of the mangrove mudflats fringing the coast of northern Australia. The choice of location means that the mud is covered by the sea for a few days only around the highest tides – the bimonthly spring tides. The rest of the time, the mud is baked hard and the crabs wait it out in their underground burrows. The only time the crabs have for feeding or mating is in the few hours after the tide recedes for a few days each month.

As soon as the mud is exposed, all of the fiddlers emerge for a frantic couple of hours of feeding before the mud dries out and sets solid. They scoop up the wet mud with their tweezer-like claws and filter it through their mouths to extract the algae and other nutrients. Though less than 6cm (2.4 inches) wide, the blue and black males are very obvious, sporting red and white claws – a flamboyant livery in stark contrast to the mud-coloured females. A male's other distinguishing feature is that he has one huge jousting claw and one small one that he feeds with. By contrast, a female can use both claws alternately to convey food to her mouth, allowing her to spend half the time that a male spends on the surface. So the mudflats end up being dominated by Technicolor males feeding away urgently.

Once refuelled, the male must perform other duties. He has to carry out running repairs on his burrow, removing unwanted sand with his legs and checking it for size with his big claw – generally sprucing it up for a potential female visitor. Then he has to set about attracting a mate. Moving around the mudflat, he raises his claw high above his head and drops it again – a combination of a white flash above his black and blue back that should prove unmissable and irresistible to any passing females.

The problem is that all that flashing also attracts the attention of predatory birds such as kingfishers, terns and kites, all of whom enjoy a meal of crab.

A mangrove kingfisher will launch itself from a vantage point in a nearby tree and then dive into the colony to grab a flasher, the shelled evidence of its success piling up beneath its perch. Kites and terns are more opportunistic, gliding over the colonies and picking up crabs that are too slow to find refuge or too far from their burrows. As a bird attacks, crabs scuttle in all directions, heading underground. It's a life-or-death game of 'musical burrows'. Such danger explains why these fiddler crabs prefer to live in colonies in the open, away from overhanging trees where predators can perch, even though that means being exposed to the scorching tropical sun. The best burrows are towards the centre, where advance warning of an aerial assault is given by the wave of panic from neighbouring crabs running for cover.

Flashing also attracts other males, who would like to expropriate the suitors' real estate, both for their own safety and because, if it is in a good location, it will be favoured by the females. So a male with a desirable residence is kept particularly busy repelling trespassers. The huge claw is pointed towards the intruder, and size definitely does matter. If the rival has a smaller claw, he retreats, but if he has a similar-sized weapon, a joust ensues. The two males lock pincers, the owner bracing himself against the lip of his burrow, and they push back and forth, trying to unbalance each other. A bout can last several minutes if the males are evenly matched, but it is usually the challenger who loses his footing and is wrenched onto his back. At this point he retires, leaving the owner guarding his burrow.

As the sun moves overhead and the temperature reaches 35°C (95°F), the owner shores up the rim and carries nutritious balls of sediment down the hole to provide refreshment for his upcoming confinement. A well-stocked pantry also makes his home a more attractive love nest for a prospective mate.

When his stalk-like eyes spot a female in range, his rate of waving increases from about five to more than twenty times per minute, and he runs over to meet her. He presents his Technicolor back to her and signals

above it with his huge white claw, while backing against her, pushing her in the direction of his burrow and vibrating his body seductively. If he is too eager and frightens her, she will duck under his claws and scuttle away. If his technique is right, she will allow herself to be propelled all the way to the burrow.

As the couple approaches the entrance, a rival male may try to intervene and steal the female. That will mean another jousting match, leaving the female to wander off. But once the resident male has lifted his rival off his feet and flung him away, he will set off to find the female again and direct her back to her new home. This time she is likely to respond by heading straight down the hole. When she doesn't emerge again, presumably satisfied with her inspection of the accommodation and the in-burrow catering, her suitor makes a final, seemingly valedictory wave and then collects a plug of mud for the burrow door. He encloses it in a 'basket' formed by the legs on one side of his body and walks back to his burrow using the legs on the other side. He then pulls the plug over his head to seal the burrow. Safe from predators and the tropical sun, the male fertilizes his mate's eggs, and the pair remain inside their temporary home for ten days or so while the eggs mature underneath the female's body.

As soon as the next spring tide washes over the mud, the mother will break out of the burrow and release her eggs into the sea to start a new generation of fiddlers.

◀ **THE JOUSTING MATCH**
Two male fiddler crabs with evenly matched claws fighting. As they struggle to and fro, the object of their passion may lose interest and wander away.

▲ **CLOSING THE HATCH**
A male sealing his burrow against the incoming tide. His larder has been stocked, and he has already led a female down the burrow.

Architect, performer, seducer

The flame bowerbird plays his part on the stage he built

Many birds use colourful feathers and dramatic displays to attract mates, and a few construct impressive, seductive bowers of twigs and leaves. The flame bowerbird does both, in style. But despite his exceptional architectural and dancing ability, this is a labour of love where anything and everything can go wrong for the suitor.

An inhabitant of the western forests of New Guinea, the male flame bowerbird's stunning yellow and red plumage originally led people to think it was a bird of paradise, but later it became the first member of the bowerbird family to be described. Most bowerbirds tend to have plain-coloured feathers and put all their effort, and often colour, into their bowers – constructions that are used simply to demonstrate what a fit and artistic male the architect is. Typically these bowers take the form of an avenue between two rows of sticks planted firmly into the ground and leading to a bower decorated with red, blue or purple leaves and berries. Some bowerbirds love these colours so much that they will collect plastic bottle tops, clothes pegs, pens and other man-made trophies (see page 79). But the flame bowerbird dazzles with his performance as much as his craftsmanship.

During the breeding season from April to September, the male builds a relatively simple bower of twigs to which he applies a fresh coat of mud paint every day. It seems that sunny weather encourages him to build, as he prefers full sunlight to illuminate his display to best effect. But he needs to find blue-coloured objects to decorate the bower. Leaves and berries make the best bouquets.

Leaving the bower unattended is risky. While he is off on a collecting mission, many things can go wrong. A female bird of paradise may raid the bower and steal sticks for her nest. A juvenile male bowerbird, come to learn from his elder and practise his building technique, may try to improve on the bower, rearranging the twigs in a different pattern. The owner then returns to a scene of devastation.

Once his bower is completed and the sun is shining, a male starts to call – a plaintive single mew. But as soon as a drab, olive-coloured female approaches, he switches to a weird electronic wheezing, which intensifies as she gets closer. While she checks out

BEAUTY IN BOTH BIRD AND BOWER
A male flame bowerbird standing in his newly decorated bower. He has the looks of a bird of paradise and the design skills of a bowerbird and must combine the two to maximum effect if he is to win what must be one of the most choosy of females.

BOWER BUILDING
Top left An immature male flame bowerbird posing next to his amateur bower and holding a beetle – a shiny love-token for a female. ***Top right*** A mature male trying to repair his bower, vandalized by a rival male.

THE ELECTRONIC-WHEEZE STAGE
With the female where he wants her – standing in the entrance of the bower, which keeps her attention on him – the male is starting the next stage of his performance. A fast rhythm of wheezing calls keeps her focused as he prepares to unfurl his plumage while holding a blue-berry love-token.

the bower at both ends, he begins to work himself up into his passionate display. He starts by shaking his vivid yellow and black wing and gradually unfurls it. He coyly bows his head below his outstretched wing, like a brightly uniformed matador holding out his cape. Slowly he rises up on his toes and then quickly jerks back down again. He repeats the action but adds a slow-motion rolling twist of his body behind the cape.

But the most impressive of displays can be ruined by the arrival of a rival male. Disturbed, both the male and female may fly off, leaving the rival to trash the bower. When the owner returns he has to rebuild the bower, painstakingly repainting it with mud. Once the sun lights it up, and as soon as a female arrives, he will resume his displaying. At the most passionate stage he will offer her a token of his esteem. If she is unimpressed by the offering, say a mauve-brown leaf, she will fly off, forcing him to up his game even further and find some vivid-blue berries to decorate the bower with.

If the female pays a return visit, he will pluck the brightest berry from the pile and resume his display, the blue of the love token contrasting with the magnificence of his plumage. Once mesmerized by his display, the female will allow him to buffet her body with his head and then mate with her. This is the suitor's only contribution to fatherhood. The female bowerbird is left to rear the young entirely on her own, and the male continues to tend his bower in the hope of attracting more lovers.

He coyly bows his head below his outstretched vivid yellow and black wing, like a brightly uniformed matador holding out his cape

Why a tiny male creates a big-wheel masterpiece

The puffer's masculinity hinges on his architectural prowess

Sometimes the more insignificant you are the more you have to do to attract a mate. It's certainly true of a drab little creature that has constructed one of nature's most intriguing valentines – a piece of handiwork that was discovered more than 15 years before the builder himself was identified.

Back in 1995, strange geometric patterns were found on the sandy seabed off the coast of the Japanese island of Amami, part of the Ryukyu chain of subtropical islands that lie to the south of mainland Japan. Local divers first noticed these mystery circles and debated whether they were some kind of natural phenomenon or the creation of a secret constructor.

Each was about a metre in diameter with a smooth inner circle of sand and about 30 channels radiating out from it like the spokes of a wheel – a perfectly symmetrical and aesthetically pleasing formation. So who or what was the mystery architect?

It wasn't until 2011 that a tiny brown pufferfish, no more than 12cm (4.7 inches) long, was finally caught in the act. And why does he do it? Because that's what a female pufferfish requires of him. She needs a perfect home for her eggs, and she will judge the male's fitness to mate by the quality of his workmanship.

He starts off his labour of love by selecting an area of sandy seabed at a depth of 10–30 metres (32–98 feet). For the first couple of days he burrows back and forth in the sand, using his pectoral and tail fins to dig channels. From the centre he creates the spokes in the wheel, stirring the sand up as he goes and leaving a series of peaks around the outer ring. But in the central hub he behaves differently.

He picks the hub clean of weed, shells and coral fragments, depositing these on the outer peaks. His digging activity and the current through the channels cause the finest sand particles to drift in clouds towards the middle. And within the hub he tills the flat surface with his specially modified anal fin, which he drags through the sand like a plough to break up the particles even further.

The whole process takes four to seven days, during which time he also chases other fish out of the area. His final act is to create an irregular wavy pattern

CONSTRUCTION
A male pufferfish pushing his body through the sand, his pectoral fins whirling the sand away, as he creates the channels that form the spokes in his wheel.

THE WHEEL OF LOVE
To create such a symmetrical masterpiece takes up to seven days. The irregular wavy lines in the middle are a sign to the female that the nest has just the right sort of soft sand in the centre, ready for her eggs.

INSPECTION AND APPROVAL
The heavily pregnant female pufferfish (left) hovering over the smooth sand in the centre, checking out its suitability as a nest for her eggs.

LOVE BITES
The male inflicting a love bite. This clinches the courtship, and the microscopic eggs are now released and fertilized.

of lines through the central hub. Now the masterpiece is ready for viewing.

The female, her belly distended with eggs, arrives to inspect his craftsmanship. But she is only allowed a quick look before he chases her away. By dawn on the following day, the wavy lines in the hub have disappeared, which seems to be the signal that everything is ready.

Now the swollen female approaches briskly and darts into the central hub. Here she hovers just above the perfectly smooth sand surface. She has chosen this male and his circle, and now she waits. He swoops down and bites her on the chin to bind their bodies close together. She shudders and sheds some eggs, which he fertilizes as they drop into the sandy nest. Over the next few hours the process is repeated until the female's belly has shrunk to normal size, while her face is bruised from his repeated love bites. When she is empty, she leaves the nest area.

Now the male switches from being a builder to a father. For the next week he tends the eggs in the hub, fanning them with his fins to oxygenate them and chasing off predatory fish. But he lets the rest of his creation be blown away in the current until only a faint impression of the pattern is left in the sand. When the eggs hatch, his parental duties are concluded. He can now become a craftsman once more, choosing a new site

and building another circle. With a breeding season that stretches from April to August, the likelihood is that he will make many of these huge constructions in a year.

But there are a couple of other mysteries still to be solved. First, if it takes about seven days of hard work to build the nest, why does the male let it become derelict and move to a new site to start again? The answer could lie in the arrangement of peaks and valleys around the central hub.

Fluid-dynamics experiments have demonstrated that as the fish burrows through the sand, the peaks funnel all the lightest and smallest sand particles towards the middle, whichever way the current is flowing. They also reduce the current's flow in the hub by 25 per cent. So the finest possible sand is blown here, accumulating to provide the best possible bed for the eggs. But once the male stops building and starts to care for the eggs, the valuable fine grains gradually get blown away, and so for his next nest he needs to find a fresh area of sand to process.

The final mystery is the precise identity of the architect himself, known only from two bays at the south of Amami island. While he clearly belongs to the pufferfish family, his looks and behaviour are like no other, and he now awaits the bestowing of a proper scientific name to complete his fame.

He swoops down and bites her on the chin to bind their bodies close together. She shudders and sheds some eggs, which he fertilizes as they drop into the sandy nest

The elegant manakin and his best man

They both have the plumage. They both do the dance. But only one gets chosen

THE MASTER
The male long-tailed manakin, attired for display. His long, trailing tail feathers will add drama to his dancing.

THE DANCING DUO
A master and his wingman on their display perch. It's the synchronicity and elegance of their combined performance that the female will judge. But even if they get ten out of ten, only the master will get to mate.

Normally males are deadly rivals in the mating game. But just occasionally they will cooperate to have the best chance of winning. The dance of the long-tailed manakin is a rare example of one such 'gentleman's agreement', where two or more males work as a team to impress potential mates.

The long-tailed manakin lives in the dense tropical forest of Central America. The male is brightly coloured, with a pair of modified streaming tail feathers that are considerably longer than his body. The female, by contrast, is a drab olive green.

A young male starts off looking like a female and goes through a series of moults each year to become increasingly brightly coloured. First the red cap, then a black face, then some black on the body before the final transformation in his fifth year into the jet-black adult with blue wings and a striking red crown. Manakins are long-lived birds, which is fortunate, since males need time both to develop their plumage to its full Technicolor glory and to hone their dancing skills.

In April the males assemble in the forest understorey to display in groups, known as leks. Each lek consists of an alpha and a beta male, the only males who can dance for females, and up to 11 lower-ranking males – the apprentices. The area around a central horizontal dancing perch is cleared of leaves and overhanging vegetation to provide the visiting females with an unrestricted view of the males' performances within the court.

Youngsters gather at their own practice sites and then dance together, but sometimes they also visit courts owned by the mature masters. Surprisingly, the young males are not attacked and are allowed to observe the proceedings. Their drab plumage appears to signal to the adult males that they are no threat. So they can watch and learn. And it's a performance worth seeing.

The alpha and beta males form alliances with each other that last up to a decade. The alpha male may be nine or ten years old, and the beta male – his wingman – a year or two younger, and they will have practised together over many years. When the alpha wants to attract a female he makes a call that sounds like 'teamoo' to enlist

▲ **DANCE AUDIENCE**
A female manakin judging the duo's performance from her perch on the display branch. One male is crouched down, running on the spot, while the other is fluttering down after a spectacular dance hop.

▶ **THE FINAL ACT**
His red crest raised, the master mates with the female, who has judged his performance to be good enough. The wingman may never get to mate, however good his dancing, unless he inherits his master's perch.

the help of his wingman. The two of them then duet in the canopy with a characteristic call that sounds so much like 'toledo' that they are known locally as toledo birds.

When the female arrives, the three of them fly down to the display perch together, and the show, which involves as many as 16 different moves, can begin. First a male will crouch and then appear to run on the spot. Then both males take it in turn to hop into the air and flutter back down, alternating the moves in synchrony, and bow to the female. As the dance intensifies, they sidle along the perch and leapfrog over each other's backs, always directing their faces towards the audience.

The female hops from one side of the branch to the other so she can observe the performance from both sides. It's a frenzy of flashing red and blue colours set off against the green forest background. Other moves include tail flicks and 180-degree turns on the perch. But they also make forays some distance from the perch to fly with laboured wing-beats like a butterfly, flitting through the understorey or back and forth from the canopy. The female is judging the performance and synchronicity of the dance to choose the best-quality male to mate with, and it may take a long time before she makes her decision. The final act is a bow by the alpha male, who raises his red crest to the female. He then dismisses his wingman with another specific call and mates with the female.

The alpha male may be nine or ten years old, and the beta male – his wingman – a year or two younger, and they will have practised together over many years

But the females aren't easy to please. If the males miss a beat, bump into each other or otherwise disrupt the synchrony of the dance, she will fly away even after watching the performance for more than 20 minutes. So the dancers have to be fit to perform – and they may have to do it to as many as ten choosy females a day.

A successful mating is all over in seconds, and the female then leaves to rear her young alone. She has no further need of her sperm-provider, as the tropical forest offers plenty of fruit for her and her young.

But the relationship between the males is much longer-lasting, and the team may dance together for nearly a decade. So what's in it for the beta male? At first glance, not a lot. Initially beta males lose weight as they have to work so hard to be accepted by the alpha, and their mortality is greater. And betas only achieve a measly 1 per cent of the matings. So the reward probably lies in simply inheriting the master's perch when he dies and taking over as the main performer – and, if a successful dancer, becoming the main sperm-provider for the females on his patch.

The stranger in the woods

Female Japanese macaques can't resist young lovers

The winners in the mating game are not always the most obvious. Dominant males may strut their stuff, but that doesn't mean they are in complete control. And a closer look at the complexity of monkey societies proves the point.

Japanese macaques live farther north than any other non-human primates – where winter temperatures are often below freezing. So not only do they need extra-dense fur but they also need to be resourceful. In the coldest part of their range they have learnt to make use of the hot-spring baths so beloved of the Japanese. On the coast, other learnt cultural behaviours include washing their food in the sea to get the sand out of it and using sea salt to season it. Once established, such habits are handed down from mothers to their offspring. Just as fascinating is the fact that coping with Japan's northern winters has also affected their mating habits.

Unlike most monkeys and apes, which live in the food-rich tropics and can give birth throughout the year, Japanese macaques need to give birth in the mild and beneficent conditions of spring, allowing enough time for their babies to grow big enough to cope with winter. Pregnancy lasts six months. So the macaques have to mate in the autumn.

As their passions rise, so their skin blushes red until their faces match the autumn colours of the Japanese maples in the forests where they roam. In troops of up to 100 animals, at any one time there may be 30 fertile females and as few as 5 adult males, but with potential rivals on the outskirts. The result is a brief, intense season when a year's worth of intrigues, violence and passion are crammed into a few short months.

Family relationships are everything to female Japanese macaques, who never leave their troops. But the males leave as adolescents and migrate up to 30km (19 miles) to seek their fortunes in another troop. Gaining acceptance can be a long and painful process, as they are aggressively rebuffed by both resident males and females, with much barking and screaming.

During the breeding season, the young males hang around the periphery of the group, sometimes displaying by shaking branches but keeping out of the way of the troop males. These dominant males, who may have been there for a decade or

THE COLOUR OF ROMANCE
A grooming pair of female Japanese macaques, their faces as red as the autumn leaves – a sign of passions stirring within. Very soon they will set about selecting which male they want to father their young.

LEADER ON ALERT
The troop leader watching for straying females. He dominates his females, and ostensibly they defer to and mate with him. But whenever his back is turned they look for younger partners .

BRANCH BRAVADO
Right A leader on top of a tree displaying his status by barking and shaking the branches. He does this especially when he has just seen off a rival male. But such bravado doesn't stop his females wandering off.

more, have established a clear hierarchy based on age, with number two giving way to number one and so on. The females defer to them, giving up the best feeding or sunbathing spots to them and, of course, mating with them.

To outsiders it must seem as if the top troop males have all they could want. But their lives aren't easy. With many females in oestrus at the same time, it can be difficult for a top male to monopolize them all. He must be constantly on the alert, surrounded as he is by younger rivals. In the morning he climbs a tree and shakes the branches, hair erect to exaggerate his size, while emitting an intimidating barking call. This he hopes will keep the outsiders at bay. But they respond with branch-shaking displays of their own.

It's not just the showy outsiders he has to worry about. A quiet rival may be employing a different strategy. Sitting in a tree he will be simply waiting to catch the eye of a female with a stare and a brief pursing of his lips – a sort of kissing action. If a female is interested, she will cautiously follow him and make cooing calls towards him. He will then lead her away out of the sight of other monkeys. For the next day or so, while the female is in oestrus, she follows him around closely. Every time he stops, she hugs him around his middle and grooms him assiduously. While together they will look around anxiously, sometimes standing up in turn to better see where the other monkeys are.

If they are undisturbed, bouts of mating, hugging and grooming follow. If the female loses her lover in the forest, she will coo to him until they are reunited. Sometimes she is so desperate to keep her toyboy (she may be twice his age) that she clings to him and rides piggyback until the male tires of the behaviour. But the illicit lovers must always be alert. If one of the troop's dominant males spots them, retribution will be swift.

Here's a typical scenario. The alpha male – the top macaque – spots one of his females with a lover. He charges – barking with rage, his fur bristling – driving the screaming female to make her escape up a tree. Honour satisfied, the aggressor returns

to his group, lies down and his more faithful females gather round to groom him. Once the alpha male's back is turned, the female rejoins her lover. Aware of something wrong, the alpha male gets up, stiff-legged, fur on end, and paces through the troop and then around it, peering into the forest. At last he spots the pair and hurtles towards them. The suitor sidesteps his attacker, but it's the female that is the alpha's real target. He catches her by her long fur and drags her around on the ground, delivering a bite before she breaks away screaming. This time she makes her way back to the troop limping. Her female relations appear to comfort her by gathering around and grooming her. But just as soon as she thinks it's safe, the female will be back with her lover again.

So for the dominant troop males, the breeding season is an exhausting round of displaying, chasing and fighting as they desperately keep up the appearance of being in charge. But despite their swagger and the punishment they dish out, it's the females who ultimately decide whom they will mate with.

At her peak of fertility, a mature female is more likely to form a consortship with a younger, strange male than the established troop males she has mated with in previous years. And DNA fingerprinting has shown that the top males father fewer babies than the up-and-coming lower-ranking ones. Perhaps this reduces the danger of inbreeding (father/daughter incest), or perhaps by having different fathers for their young, female macaques are doing the genetic equivalent of not putting all their eggs in one basket.

Female power doesn't end with the mating game. Males who are popular with the females are more likely to maintain their position at the top of the hierarchy even when they become weaker than younger males. And females will gang up on any male who attacks a member of their family. So it seems that, in Japanese macaque society, males may appear to be in charge, but it's ultimately the females who pull the strings.

▲ **CHARGE OF THE ALPHA MALE**
Left Having spotted a rival male with one of his females, the alpha male races towards the pair, fur erect. More often than not he vents his displeasure on the female rather than his rival.

▲ **SNEAKY COUPLE**
A sneaky pair looking nervously up the hill in case the alpha male has spotted them. The female has chosen the most fertile phase of her cycle, when she is most likely to become pregnant, to conduct this illicit romance.

▼ **SOCIAL HUDDLE**
Overleaf Huddles of females preparing for a cold night. It's the females who make long-lasting relationships that bind the troop together. They not only choose who they mate with but they also influence who stays as the troop's alpha male.

Dangerous dancing

When style and speed make the difference between procreation and death

Males devote huge amounts of time and energy in their efforts to attract a female. But rarely are the stakes as high as those for the peacock spiders of southeastern Australia. For these tiny males, failure to impress can be fatal.

Peacock spiders belong to the family known as jumping spiders, a highly charismatic group, with extraordinary lifestyles and huge personalities packed into tiny bodies. Though only 5mm long, both sexes of peacock spiders are voracious hunters, tackling insects much larger than themselves. They have large eyes for daylight hunting and can jump many times their own body length.

When a female peacock spider is receptive, she lays a line of silk impregnated with a pheromone (sexual scent) for males to follow through the undergrowth. But even when he knows she is interested in receiving his sperm, he has to take care, because as soon as she

SIGNING DESIRE
Waving his long, white-tipped legs on alternate sides of his body and high above his head, a male peacock spider is semaphoring his good intentions to a prospective mate.

FAN-DANCING SHOW
Opposite page The female is interested in the male, who is stepping up his performance by unfurling his Technicolor fan between his upraised legs. He enhances the vivid pattern by scuttling to and fro in front of her to create a moving kaleidoscope of colour.

has spotted him, he needs her to know that he would make a better suitor than a meal. She is brown, but he sports an iridescent red and blue coat. The colours are obviously attractive to her, but they aren't enough. He also has to signal his amorous intentions.

Now he starts bobbing his abdomen to produce vibrations that the female can sense as well as see. His front appendages, his pedipalps, also flicker with excitement. They hold packages of his sperm for her. Getting closer, he then uses a form of semaphore with his extra-long third pair of legs. Covered with black hairs, tipped by a distinctive white tuft, they function as flags. Raising both legs, he waves them simultaneously or alternately over his head and approaches, scuttling sideways to intensify the effect. Moving in a semicircle around the female, going in one direction and then back, he gets closer and closer to her. She turns to get a better view. But all of this so far still isn't enough. Now he has to dance for her.

He unfurls a multicoloured fan like a peacock's tail from around his abdomen and prances back and forth in front of her. For maximum effect, he synchronizes all his moves – the leg waves, the fan flickering and the bobbing – in one colourful dance. He pauses only to observe the effect on her and to check she's paying attention. Then he lets the fan drop and flutters it provocatively. This does the trick, and she swivels again to get a better look.

Now comes the finale. He furls the fan back around his abdomen, spreads his third pair of legs out wide and extends his first pair of legs towards her while vibrating his body. It is a truly spectacular performance, and he is rewarded when the female allows him to insert his pedipalps into her genitalia and deposit the sperm. But as soon as he has removed them, he has to flee. In the female's eyes, the male is now a meal rather than a mate and she would happily have her cake and eat it, too.

FATAL ATTRACTION
The male performing very close to the female. If his show is top rate, she will resist the urge to eat him and allow him close enough to deposit his sperm. But after mating he must move fast to avoid becoming her post-coital dinner.

Parenthood

To be a parent is the final mark of success in any animal's life. It means that animal has passed on its own genes to a new generation, achieving the next-best thing to immortality. But for different creatures, parenthood involves different responsibilities – from the insects or fish whose duties are over once the season's eggs are laid, to a mother bonobo who will watch over her son for the rest of her life.

THE EXTENDED FAMILY
Meerkats warming up together. This is one big extended family as far as the youngsters are concerned. Everyone helps out with the parenting. Aunts, uncles and older siblings will all bring food or care for them while the rest of the group is out foraging.

BABY BOY BONOBO
Previous page A young male lying on his devoted mother. She will continue to suckle him until he is four or five years old.

Parenthood

At its simplest, parenting may mean no more than choosing a safe place to deposit fertilized eggs: a salmon's scrape in a riverbed or, for a turtle, a nest in warm sand. The hardest-working parent tends to be the female. Eggs are bigger than sperm, and so from the start the mother's investment is greater. This is especially true of birds and mammals. Birds produce relatively large, shelled eggs, and mammals not only have to undergo pregnancy but also lactation, which involves a huge transfer of energy and nutrients from the mother to her young. So inevitably, because she has invested so much, it's the mother who is left holding the baby. But there are exceptions. Male water bugs and midwife toads, for example, carry their eggs until they hatch, and the male ruff incubates his clutch and cares for the chicks on his own, allowing the female to look for more fathers to care for a new clutch of eggs and young.

Parenting duties are, though, shared by 90 per cent of male birds, who can feed and incubate chicks as well as a female can. This is not true of mammals – only the mother can nurse the young – and so relatively few fathers become heavily involved in childcare. These include wild dogs and wolves and mongooses, where the father provides food for the family, and marmoset and tamarin monkeys, where fathers carry their twin young, who visit their mothers only at feeding times. Fathers of Mongolian gerbils retrieve, lick and huddle over their young, behaviour brought on by pheromones (hormones) from their pregnant partners.

But parenting duties may also be shared beyond mothers and fathers. Eider ducklings form crèches of 100 or more, often watched over by 'aunties' – females who have failed to breed or lost their clutches. In social monkeys and apes, siblings, aunts and grandmothers all share in the care of the babies, and the broader family care of elephants has been shown to be crucial, especially for first-time mothers.

Sometimes parental drive can have unfortunate consequences. A dominant female meerkat will viciously oust any of her pregnant daughters, to ensure that only her latest offspring get the undivided help of the troop. Female monkeys and apes sometimes kidnap the babies of subordinate females. And some birds leave their eggs in the nests of other birds for them to rear. For example, many female ducks, while raising clutches of their own, often lay extra eggs in nests of other ducks. As most ducks are ground-nesters and lose many clutches to predators, this is a way of not putting all your eggs in one basket.

The most dramatic form of parenting is to risk your life for your offspring. When a predator approaches a plover's nest or young, the parent attracts attention to itself by behaving as if it's injured, fluttering just out of reach of the hunters jaws, leading it away. A female cheetah will defend her cubs against a much larger lion, and a moose will rain blows with its hooves on a wolf or bear in defence of its calf.

But the ultimate maternal sacrifice is found among the arachnids (spiders and the like). Some mother spiders and pseudoscorpions, which care for their babies for weeks after hatching, encourage their young to eat them, either when their job as a parent is over or in response to a lack of food. They signal their last act of motherly love by drumming or by raising their limbs in a maternal gesture, and then the young swarm all over them to begin their feast. This reduces cannibalism within the brood and is a successful strategy once a mother's work is finally done.

ULTIMATE DEVOTION
A moose defending her newborn calf from a wolf pack in a tundra pond in Denali National Park, Alaska. Driving down on them with her powerful front hooves, she manages to fend them off for 15 minutes, but in the end, the wolves get their meal. Later, a trapper killed the mother wolf – the family's key hunter.

The greatest sacrifice of all

The tale of Raine Island is one of determination and danger

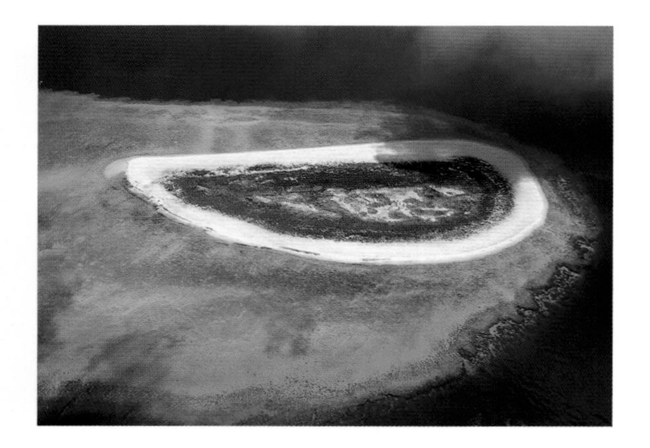

ISLAND LURE
Raine Island, at the northerly tip of the Great Barrier Reef. It's 32ha (79 acres) in area, set within a wider underwater reef, and its coralline sandy beaches are a magnet for female green turtles.

THE GREAT MIGRATION
Thousands of green turtles swimming to the island to lay their eggs. Gathering off the edge of the reef that fringes the island, they will wait for the relative cool of night before crawling ashore.

Mother turtles, like most other marine reptiles, have to come ashore to lay their eggs. Though they don't look after their babies when they hatch, they go to extremes to give them the best start in life. This means finding the perfect place to bury their eggs – a nest on a dry, sandy beach with full sun for incubation and high enough up to avoid tidal flooding. For an animal with a body adapted to marine life, the land is a hostile place. Gravity kicks in as a mother hauls her 150kg (330-pound) body up the beach with her now-ineffective flippers. She may also be in competition with thousands of other turtles, all after the same thing. This can lead to disaster.

One of the most important nesting grounds for green turtles is the coral cay of Raine Island, on the northernmost end of the Great Barrier Reef. Visited by females from all around northern Australia, Indonesia and Papua New Guinea, as many as 10,000 may nest on Raine in a single night. Raine only became an island about 4,000 years ago, when the sea-levels dropped, but there is fossil evidence of nesting by green turtles here for more than a thousand years, making it the longest known continually used nesting site for green turtles.

It is treeless and uninhabited except by seabirds. It is also the site of what nineteenth-century Europeans believed to be a turtle graveyard. But the reason for the masses of shells, bones and carcasses is more gruesome.

The turtles move ashore on a summer night onto the white coral sand that rings the island – the same place where they emerged as hatchlings perhaps 50 years ago. They have fewer than 12 hours before the sun rises and bakes the beaches again, but the area of sand untouched by high tides extends for only about 30–90 metres (100–300 feet).

On a busy night, up to 10,000 turtles struggle into this zone. Inevitably, as one digs its metre-deep pit to lay its clutch, another is alongside, flipping sand in its face and even bumping into it. Many mothers head inland. Some may spend up to six hours looking for a peaceful spot to lay in, before giving up and returning to the sea. They will try again on subsequent nights and if forced to give up will reabsorb their 100 or so eggs. As females visit the nesting beaches only once every five years, this

Visited by females from all around Northern Australia, Indonesia and Papua New Guinea, as many as 10,000 may nest on Raine in a single night

HEADING FOR THE BIRTHING BEACH
Green turtles, heavy with fertilized eggs, on their way back to Raine Island, where they themselves were born. Perfectly adapted to life in the water, they have highly tuned navigational abilities. But to lay their eggs, these females will have to leave the cool safety of the sea and drag themselves onto the island, a harsh, alien and dangerous world.

is a significant loss of future hatchlings. But these turtles may be the lucky ones.

The turtles that head into the island's central depression, which is mainly rocky and unsuitable for nesting, may be lucky and find a rare sandy spot to lay their clutch. But their troubles are only beginning.

Turtles navigate their way back to the water by looking for a bright area lower down on the horizon, which if they are at the top of a beach works well. But within a sunken area, they are reluctant to head uphill. They spend hours crisscrossing the centre of the island until, as the sun comes up, they start to suffer from heat exhaustion.

Hours of direct tropical sun on their dark backs, which absorb the heat rapidly, can mean they bake to death in their shells by early afternoon. Of the turtles that manage to head back towards the beach, many will flip over as they fall over the rocks at one end of the island and lie helpless on their backs. They can survive in this position for days because their pale under-shells absorb much less heat, but eventually the result is the same. Even those that manage to get back down to the seashore can be trapped at low tide by exposed reefs. But the death of so many animals is a boon for others. Turtle carcasses washed out to sea on high tides attract top predators. Tiger sharks are drawn to the islands during the nesting season to take advantage of this annual feast. Mostly they scavenge, but exhausted turtles, returning after their arduous assault course, are tempting targets, and some receive wounds or lose flippers before reaching safety in the open ocean.

With all the horrors these turtles face in their attempts to nest, it's even more impressive that they keep on coming and that so many do succeed.

Around 2,000 turtles die on the island during a big nesting season, and though the process is a natural one, this population is too important for scientists not to try to help them. Since the 1980s scientists have been righting any upended turtles they find and helping them find their way back to the sea. More recently fences have been put up to keep the turtles out of the most dangerous areas. And plans are afoot to dump extra sand on the island to provide more dry nesting areas and to provide ramps for the turtles to navigate the uneven terrain more easily. So the extraordinary spectacle of thousands of turtles doggedly invading the beaches as they risk their lives to become mothers will continue.

TRAPPED IN THE SUN
A female trapped in the rocks on the way back to the sea. She faces death within a matter of hours of exposure to the tropical sun. Her dark shell absorbs heat rapidly, and she has no means of losing heat by panting or sweating.

FREED BY THE TIDE
A female trapped on the outer edge of the reef in the tidal zone. Here the danger is not from overheating in the sun but from drowning in the incoming tide. But in this case, the inrushing sea gave the turtle the lift it needed to float free.

Conning the drongos

The parents that are duped into raising an imposter

To rear a family successfully requires unconditional parental commitment. It's hard work creating a home and finding food for a growing brood, but for some birds that blind devotion can have unfortunate consequences. The fork-tailed drongo is a familiar sight in southern Africa, perched watching for insects. These pugnacious birds are capable of defending themselves against predators and regularly mob hawks, owls and even snakes and mongooses. But they can be outwitted by a master of deceit.

Drongos build their nests each spring in spindly forks at the tips of tree branches in savannah woodland. When the female is ready to lay her eggs, the pair have to be vigilant, because the nest may be staked out by a female African cuckoo. Like cuckoos anywhere, all her energy is devoted to duping someone else into bringing up her baby. And for the African cuckoo, that unlucky someone is always a fork-tailed drongo.

The cuckoo checks out all the nests in the area and monitors their progress. She will only lay her perfect replica of a drongo egg – exactly the same size and complete

THE LITTLE EGG MOVER
An African cuckoo chick evicting an egg from a drongo's nest. Cuckoos' eggs develop faster than their hosts', giving the cuckoo chick a head start so it can get rid of the competition.

CUCKOOED
Opposite page A drongo feeding the imposter. It is also attempting to shade it from the sun, which becomes increasingly difficult as the alien chick grows bigger and bigger.

with the right colour and markings – while the drongo is laying her eggs. A drongo knows that a cuckoo means trouble and will mob it. But over the period that the female drongo lays her eggs, one a day for three days, she has to spend some time away from the nest, feeding herself – vital if she is to nourish her developing eggs. While she is away, the male guards the nest, but to outwit him, the pair of cuckoos may even act in concert – the male distracting the drongo while the female cuckoo sneaks in. It only takes seconds for her to remove a drongo egg and replace it by laying one of her own so that nothing seems to have changed.

Individual drongos lay eggs in a variety of colours (from white to pale red to rich buff) and patterns (from unmarked to speckled to heavily blotched), and so a female cuckoo has to match the specific pattern of the other eggs so that hers won't be rejected and removed. It's a gamble. Each cuckoo can only lay one type of patterned egg. If she's lucky it matches the host's coloration. If not, the egg is rejected, and indeed many eggs are removed – pierced and hoisted over the side by a drongo's sharp beak (though the cuckoo's eggshell is also thicker than normal – perhaps to resist pecking by a suspicious host).

Being suspicious is second nature to the drongo. Sometimes a drongo returns in time to catch the intruder in action, and cuckoo feathers found below nests testify to the ferocity of the resulting attack. So the ongoing battle between cuckoo and host seems to be both a sensory and a physical one.

For the next 16 days, the drongo pair would normally take turns incubating their two or three eggs, but now that a cuckoo is in the clutch, their family is doomed. The mother cuckoo has given her chick one other advantage: she has kept the egg in her body a day longer than most other birds, allowing it to develop a little further, which means it will almost certainly hatch before the drongo eggs. Blind and weighing only a few grams, the cuckoo chick then performs a Herculean task. It manoeuvres its body underneath each egg and braces itself, arching its concave back to form a cup in which the egg is pushed up and over the edge of the nest. The cuckoo rests between evictions, gathering its strength for the next removal until all eggs are gone. Now it has the undivided attention of both parents as it eats and eats, enough for three.

The chick's intense begging invokes an irresistible urge in the drongos to fill its gaping red maw. And when it's hot, they shade the chick under their spread wings until it can barely fit beneath them. After three weeks, the monster cuckoo is too big for the nest. But the drongos continue to feed it, even after it has fledged. And after eight weeks of wasted effort on the interloper, they have lost the chance to breed again and must wait a whole year before they can try once more for true parenthood.

The chick's intense begging invokes an irresistible urge in the drongos to fill its gaping red maw. And when it's hot, they shade the chick under their spread wings until it can barely fit beneath them

EGG SHIFTING
Top left An African cuckoo hatchling prepares to heave out the eggs that would hatch into potential rivals for food – those laid by its new parents. ***Top right and bottom right*** Two clutches of fork-tailed drongo eggs – each including, on the right, an African cuckoo egg. This shows just how good a match a cuckoo egg needs to be to avoid being chucked out by the drongo parents. ***Bottom left*** A successful cuckoo chick, just a few days old and now sole occupier of the drongo nest.

Defender of the family

In langur society, everyone has a hand bringing up baby, but father's defence of his offspring is crucial

In the highly social world of monkeys, parenting can take many forms, and it's not always just the mother's responsibility. In Hanuman langurs everyone seems to want to get in on the act.

The mother of course is the main carer but other females and younger monkeys will also help, given the chance. And though the father appears to ignore his offspring most of the time, he has a vital role to play in ensuring their safety.

The best-studied Hanuman langurs are those living in and around the Indian city of Jodhpur. Named after the Hindu monkey god, they are considered sacred. People feed them and tolerate their raids on farms and orchards. With plenty of food and water, the monkeys are also plentiful. A typical Jodhpur troop has just one male and many females and can number more than a hundred animals. There are also bachelor groups of young males.

Though the maternal bond is very strong – mothers have been seen to carry dead infants around with them for several weeks – a mother will also allow other females to carry her baby and groom and play with it. This may free her to feed unencumbered. But less experienced babysitters can be a problem.

Young langurs are fascinated by babies and will queue up beside a mother to touch and, ideally, carry the infant. Initially it appears to enjoy the attention, but one of the favourite games in a group of youngsters appears to be 'pass the baby', which may result in an infant being held upside down, sat on, dragged along the ground or dropped from a height. So the mother always has to be on the lookout for bullying or accidents in this school of hard knocks and be ready to come to the rescue of her screaming infant.

The father tolerates his babies, and they seem pleased to see him when he returns from patrolling the edge of the troop's territory. But it's only when a bachelor band challenges his supremacy that his true value as a parent comes into play.

Young bachelor males have been evicted from their natal troops and are now looking to break into new ones. They will harass a troop leader, hoping to overthrow

LEADER OF THE TROOP
The alpha male of his troop. Though he does little in the way of parenting – it's the females who suckle and care for their young – he has a hero's role to play.

him and then mate with the females. But a female isn't receptive (doesn't come into oestrus) while she is nursing an infant, which she continues to do until it is a year old. So the first act of any successful bachelor group's leader, after overthrowing an established male, is to attempt to kill all the babies under a year old.

The resident male will have sired most, if not all, the young in his troop – they are his legacy. So when the bachelors get too close, he whoops and bounds around his territory displaying his strength and fitness. If that doesn't work, a bloody battle ensues that can result in horrendous injuries or even death.

So when the resident male successfully sees off his rivals and returns to lick his wounds, his offspring gather around in an excited welcome.

They should indeed be grateful because he has saved their lives, and his actions mean they are one step closer to surviving to adulthood.

MOTHER-BABY GATHERING
Part of a Jodhpur troop of langur mothers, all with their infants. They make doting mothers and will suckle their offspring until they are at least a year old and have their adult coloration. But they have much to fear from a maurading gang of young males.

The resident male will have sired most, if not all, the young in his troop – they are his legacy. So when the bachelors get too close, he whoops and bounds around his territory displaying his strength and fitness

SCATTERING THE BAD BOYS
An alpha male troop leader attacks a menacing group of bachelor males, showing what he is made of as a father. Given the chance, the bachelors would kill his offspring, and so it is the youngsters' survival that he is fighting for.

The bad babies

A cattle egret nest turns into a scene of jealousy, mayhem and murder

THE BIGGEST GETS THE BEAKFUL
A parent egret, its crop full of food to regurgitate to its chicks, struggling to find a way to feed them. Just one – the first out of the egg and the biggest – is most likely to get the beakful. As the cattle egret chicks grow and fighting becomes more vicious, feeding time becomes ever more difficult.

Cattle egrets are the most abundant of the heron family and one of the great bird success stories. Since the late 1800s they have spread from their original stronghold in Africa, where they foraged alongside zebras and buffalo, and now can be found worldwide. By the middle of the twentieth century, they had crossed the Atlantic to North America and are now common in the swamps of Florida and Louisiana, where they breed alongside other herons. As parents, they make a hard-working team, sharing duties evenly, and they are endlessly resilient in meeting the demands of their young.

The cattle egrets nest alongside other herons, egrets and spoonbills on islands in lakes and swamps where they are safer from terrestrial predators. There is intense competition for the best positions and nesting material. And as the nests are often built in low trees over water, with alligators loitering below, a secure platform is essential to support the weight of both adults and growing chicks. It's the male that chooses the location, normally in the fork of a branch. There he displays his orange plumes to attract a partner and fend off other prospective homebuilders. Once he secures a mate, she starts building the nest while he collects sticks from the ground or from his neighbours. If he finds a nest and is caught stealing, a skirmish of stabbing beaks ensues.

On the first day of nest-building, the sticks he has brought, one at a time, are simply laid across a tree fork, with many dropping to the floor. But by the second day, some are wedged into place, and the construction gradually starts to take shape. The female inserts new twigs into the pile to strengthen it, and eventually both parents work together to tidy up loose ends. After six days the nest is ready for egg-laying – though stick-collecting and building work continues even after the chicks have hatched.

The female lays an egg every couple of days until she has a clutch of three or four. Then the three weeks of incubation is shared equally by both parents. Hatching is staggered, and there is a significant size difference between the first and the last arrivals. This means the first born is stronger and bigger and gets more food than the second and so on. Within a day of hatching, the first chick begins pecking at its parent's bill – a signal for the parent to regurgitate a meal of partially digested

insects – and in the next few days the younger siblings, as they hatch, join in. But at five or six days old, the eldest chick is strong enough to grasp the tip of the parent's bill and swallow the whole meal itself. The younger ones are now only able to feed once the eldest is full, and this reinforces the siblings' disparity in size.

One parent broods the family constantly for the first two weeks until the chicks are well-enough feathered to maintain body temperature. From now on the parents' visits to the nest become more and more of a trial, as the young attempt to seize the adult's bill and win the prize, bobbing their heads, waving their wings and quivering with excitement. Between feeds the parents hang back out of reach of the squabbling youngsters and do nothing to correct their unruly behaviour. When food is scarce, they will even stand by and watch as the youngest chick is bullied until it ceases to compete for food and starves. It seems heartless, but it's better to rear fewer, stronger chicks than to risk starving the whole family.

By four weeks the youngsters have left the nest to perch in the tree. Now their begging reaches a peak, and they will chase a parent along the branches, buffeting it with their wings and thrusting dagger-like beaks into its face.

The adults now spend most of their time away from their unruly brood, approaching them rarely and with less and less enthusiasm until, at about six weeks, the young are finally ready to feed themselves. After ten weeks they are independent, and the parents can resume their more sedate lives until next year.

MUGGING TIME
A clutch of nearly grown teenagers shouting for food. Their dagger-like beaks make feeding time a daunting prospect for a worn-out parent. No wonder the parents now spend as little time with them as possible.

COURTSHIP DAYS
Courtship on the nest – the start of it all. Back then, life for the parents revolved around nest-building and courtship. The male would offer his mate twigs as love tokens, and they would display their beautiful plumes, their beaks and irises suffused with orange-red.

A boy bonobo's heart will always belong to mama

And why a mother's devotion can help your social standing

When we think of the best parents in the non-human world, we often turn to humanity's nearest relative, a position now held jointly by the chimpanzee and the bonobo. The bonobo was the last of the great apes to be discovered, only identified as a species separate from the chimpanzee as late as 1927. It was originally called the pygmy chimpanzee, because of its more slender body shape rather than its overall size, and it is still the most mysterious member of the endangered family.

While best known for its apparently peaceful social life and enthusiasm for playful sex, bonobos are also remarkable as parents. The bond between mother and son is now believed to be the most important one in bonobo society. Mother bonobos are more attentive and take more concentrated care of their young than do chimps, and the mother-son relationship benefits both of them for the rest of their lives.

Bonobos are so little known because they are only found in one country, the Democratic Republic of Congo, and then just on the south side of the Congo River. One theory to explain their evolution from chimps about 900,000 years ago is the absence of gorillas. On the north bank, gorillas have specialized in feeding on herbs and vegetation, forcing chimps to harvest patchily available fruit and occasional meat to avoid competing directly with them. The bonobos, alone in the south, have been able to exploit the steady plenty of the gorilla diet, alongside the higher value but scarcer chimp diet. The combined richness and constant availability of food, without competition from the larger apes, may have also allowed the bonobos to live in larger, more stable groups than chimps and, ultimately, in a very different type of society.

Bonobos live in mixed groups of 10–30 animals, including several males, females and their young. Female bonobos leave their natal groups as adolescents at about seven years of age. To be accepted in a new group, they have to work hard on making friends with the female residents by grooming and sexual behaviour, thereby gradually rising up the female hierarchy. Unlike the other great apes, the females can

PRECIOUS BOY
A young male bonobo, still very close to mother. He will maintain a relationship with her for the rest of her life. She will reassure him when needed and support him in bonobo social politics. It's possibly the longest mother-son relationship to be found in nature.

TRAVELLING WITH MOTHER
A young male riding on his mother's back to the next foraging site. Bonobos travel as much as 20km (more than 12 miles) in a day, and so carrying the youngster becomes more and more of a chore as he grows older and heavier.

be dominant over the males, and clashes between the sexes are rare. Tensions over food or when groups reunite are often dispelled by sexual activity, in which all the animals may take part in any combination of age or sex. When a female becomes sexually receptive, males rarely intimidate or attack her as this may bring down the wrath of the whole group, including the other females. A better strategy for males is to maintain long-term friendly relationships with all the females so they can be close when mating opportunities arise. And it seems to be this that makes the mother-son relationship so important in the long term.

Baby bonobos are particularly helpless, clinging to their mother's bellies for the first three months of their lives, and don't move more than a metre or so from them until they are more than six months old. For the first year of life, they depend totally on their mother's milk and aren't weaned till they're four or five. Only in their third year do they even start to ride on their mothers' backs, which is particularly important when travelling longer distances.

A bonobo can range 10–15km in a day, and so transport of the growing infant becomes more onerous for the dedicated mother. She regularly visits swamps in remote parts of the forest to find mineral-rich water plants that seem necessary for their diet.

The longer-legged bonobos are more likely and better able than chimps to walk upright, and this ability is particularly useful for wading through the water to find lily stalks.

It's now that the infants, averse to getting wet, climb onto their mothers' shoulders on the seashore, looking for all the world like human babies being carried by their mothers, which gives us a glimpse of our own ancestral past.

From three years of age, a young bonobo begins to play at making nests in the trees, but it continues to return to sleep in mum's bed for another couple of years. All this work by the mothers pays off, as more than 80 per cent of young bonobos survive to their sixth year.

Within the wider family group, a mother is the centre of the world, and for a son she will remain so for the rest of his life. Her rank increases if she gives birth to a male, and her son's rank is positively related to hers. Mothers and adult sons associate together more than other bonobos, and the presence of his mother even influences his success at fathering sons. Her presence seems to enable him to be more readily accepted by the females around her, and she may also provide moral and physical support when he is competing with other males for a receptive female. So for success in the mating game, a bonobo's best friend is his mother.

▲ **LEARNING WITH MOTHER**
Mother sharing a junglesop fruit with her small son and then another youngster, while a male waits for a handout. Her son will learn from her not only what to eat but also about bonobo social politics and his status in it.

▼ **PLAYING WITH MOTHER**
Overleaf A youngster happy in the arms of his mother. She will be his first and his main playmate, and he will continue to share her tree nest-bed until he is five years old.

Mother's big helpers

In a herd, a young elephant is everybody's baby

BABY CARE
A very young calf being helped to its feet by the trunks of its mother and 'aunties'. As it grows up, there will always be a trunk nearby for consolation should it be in difficulty or distress or simply need reassurance.

THE MATRIARCHAL CLAN
Part of an all-female clan – mothers, sisters, daughters and even granddaughters. They work together, helping and supporting each other.

The African proverb that it takes a village to raise a child is paralleled in the lives of Africa's largest mammal, the elephant. The larger the family herd, the more females there are who can act as aunties, nurses, guards and helpers, and the more likely a new calf is to survive. This support is vital for a first-time mother.

Elephants first give birth at 13 or 14 years old and may then have a calf every 4 to 5 years for the next 50 years or so. An experienced mother is constantly attentive to her calf for the first year of its life and maintains a close bond until it's more than ten. In the case of female calves, that bond will last for life. A mother calls to and constantly caresses her newborn calf. She props it up using her trunk and will bend her front legs to lower her breast and make suckling easier. She pushes it under her body if there is danger or to shade it from the sun and, having chosen the safest place to cross a river, pushes and pulls it up and down the banks. When walking, she physically stays in constant touch with her baby via her sensitive tail, stopping when it stops and allowing it to feed. Experienced mothers are very good at determining when their baby's demands to suckle represent a genuine need, whether for comfort or nutrition.

But a first-time mother may be shocked by the new arrival and can find it difficult to respond accurately to her baby's needs. Then the reassurance given by her own mother or by other older females can be crucial. Grandmothers will help calves to cross rivers, even elbowing a young inexperienced mother out of the way to retrieve a struggling calf stuck in the mud or unable to climb a bank. Grandmothers also not infrequently suckle their grandchildren.

New mothers are less likely to wait for their young as they travel, and so their calves get lost and become distressed more often, calling and ear-flapping in alarm. But the response from other females in the wider family is swift. Nevertheless, calves of first-time mothers are twice as likely to die as those of experienced mothers. So while all calves benefit from having aunties and carers in the group, it's those of the first-time mothers that need the help the most. Perhaps because bringing up a baby elephant requires such skill and dedication, adolescent female elephants are

irresistibly drawn to the infants, caressing them with their trunks and shepherding them along in the group. They are quick to respond to distress calls, and so the young calves are never far from a reassuring trunk. Adolescents will kneel or lie down to solicit play from the youngsters – play which enables them to become more coordinated and educates the young ones in the ways of their elders.

At the other end of the age range, the matriarch is also crucial to the youngster's survival through the power of her leadership. A wise and experienced female who makes good choices about when and where to travel will increase the reproductive success of her whole family, increasing calf survival by ensuring everyone benefits from access to plentiful food and water while avoiding danger.

So, for elephants, mothering is the ultimate community affair.

▲ **GRANNY CARE**
An elderly mother watching over the play of both her own two calves and the two she has adopted, orphaned by the death of their mothers – her daughters.

◀ **GREETING THE NEWBORN**
A wobbly newborn being welcomed with excitement by members of her mother's matriarchal group.

7

Life Story's own story

In the filming of *Life Story* – 'the greatest of all adventures' – the teams have experienced surprises and hardships, matched their wits against the odds, suffered and celebrated, and come back home, tired and exhilarated, with amazing discoveries and gripping tales.

MOVING DAY
After the storm, Sweden – the end of the first Arctic fox shoot but not the end of the story. Camera gear, food store and generator were dug out and transferred to the next shoot.

INVESTIGATING A CAMERAMAN
Previous page A meerkat assessing cameraman Toby Strong as a climbing-frame from which to scan the horizon and the sky for predators.

Filming the greatest of all adventures

Life sharper than it's ever been seen

Early January in the Kalahari and the sun has been up for less than an hour. A baby meerkat, just three weeks old, has emerged from its burrow for the first time in its life. Watching and waiting just a few feet away is David Attenborough. He begins to talk of the dramatic and almost certainly perilous life this little creature is embarking upon, then wishes it 'Good luck'… There is a moment's pause before director and series producer Rupert Barrington quietly calls 'cut!'

With the passing of those last few frames through the camera, two and a half years of filming and nearly four years of production came to an end. Those two and a half years had seen the production team scour the four corners of the planet, mount 78 expeditions, spend 1,917 days in the field, shoot 1,800 hours of footage and experience countless adventures, some of which are recounted in the following pages.

The sequence with the meerkat opens the series and establishes its unique approach. As David says: 'There is a story that unites each of us with every creature on the planet. It is the story of the greatest of all adventures – the journey through life.' And that's the journey *Life Story* follows, from birth to nature's perpetual goal, the production of the next generation. The approach has been to delve deeply into the lives of the animals that have been filmed, to understand their individual challenges, and to see the choices they make and the repercussions of those choices.

Life Story has been lucky enough to be the first natural history series television shot using ultra-HD cameras. Quantum shifts in image quality are rare. The last came more than a decade ago, when high-definition cameras first appeared. The latest – the first 4K cameras (four times better resolution than HD) – arrived when the series was still in the planning stages. Though these cameras were untested in the kind of environments the series was to shoot in and the rough treatment they would be put through, they offered crucial advantages: 4K brings startling depth and detail to a shot, as well as vivid colours and textures. This would allow the characters of the animals – their reactions to success and failure, the intensity with which they approach a task, their fear and, for some, their joy – to play out on their faces. Close-up shots of animals would be central to the

BEFORE THE SHOW STARTS
Paul Stewart filming sunrise in Costa Rica before picking up his camera and setting off into the rainforest to film displaying manakins.

STICKING CLOSE TO A BULL SHARK
Cameraman Didier Noirot focusing on a remora stuck to a bull shark, waiting to see if a jackfish might eat it. The shark was not interested in eating the human.

rEvo
SUBSPACE

WATCHING FOR EGRETS – AND ALLIGATORS
Mark Payne-Gill filming nesting cattle egrets in a Louisiana swamp. The only way to get near enough was to stand in the water, very still, to avoid both disturbing the birds and attracting the alligators.

series. If filmed in ultra-HD, every scratch and crease on an animal's face would be visible, painting a sense of character in a way that had not been possible before.

Second, the large size of the sensors on these cameras would mean that the depth of focus in the image was much shallower – in other words, less of the image would be sharp. This would allow the camera operators to isolate an animal in a shallow plane of focus, drawing the viewers' attention to it, while other animals or the landscape would fall out of focus. This would also add intensity to close-up portraits where perhaps only the eyes would be sharp.

So the decision was taken to risk using them in the field. Whenever possible the cameras were to be taken off the tripod and placed down in the animals' world, and by moving the camera alongside the subjects with specially designed gyro-stabilized rigs, the camera operator would achieve a unique freedom, and the images a unique fluidity.

Of course all the technological advances in the world give you little unless you can find new and exciting stories and the animals to tell them. The most compelling tales are those in which the stakes are as high as can be, where there is everything to lose and everything to gain.

In our own lives such stories are probably rare. Most of us will never face the threat of death before our time – the threat that our lives will amount to absolutely nothing. But for many wild animals, those threats are a constant. Most animals' lives are hard for most of the time, and to survive takes astonishing tenacity, inventiveness and luck.

Understanding the animals comes from long detailed study. A recurring theme throughout our filming has been the reliance on the extraordinary band of field researchers and scientists who dedicate their lives to understanding their subjects, spending year after year

TINY BIG STAR
The peacock spider, not much bigger than a match-head, in full display. He was on a mission, oblivious of the camera, often stopping and posing, but capable of moving with sudden speed.

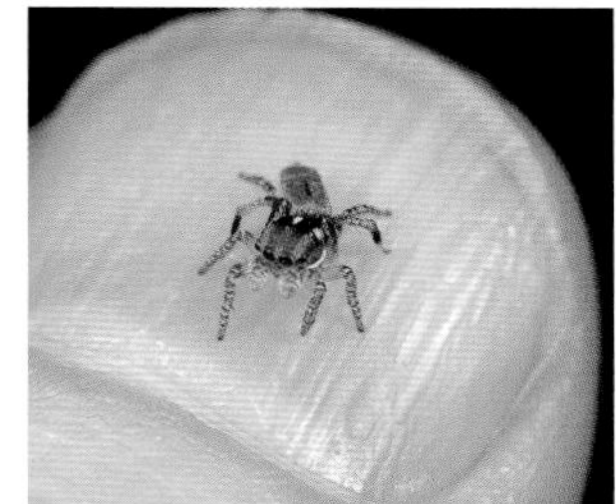

in remote, uncomfortable and sometimes alarmingly hostile places. When crews go to such places they often have to share that discomfort, but that is rarely any real hardship – if the result of the efforts is to open a window on the lives of wonderful animals.

The first test of our approach and the first shoot of the series was the dance of the peacock jumping spider. This recently discovered behaviour is spectacularly colourful. The spider is minute and notoriously unpredictable, sitting still for long periods and then, without warning, moving at lightning speed. Clearly, it was going to present a technical challenge. But first it was necessary to work out how to tell its story – how to draw the viewer into the spider's journey, which culminates with it dancing for its life.

TREETOP ANT VIEWS
A crane hoisting the team to film the weaver ants in their leaf-pod treetop homes in Queensland, Australia.

The dance itself, though the highlight, is far from the whole story, and the telling of the story would be as important as the ending. We decided to start from the time the male spider has reached maturity and begins his search for a mate and then to draw in the viewer by creating a sort of mystery, with clues and a nasty surprise.

The spider searches the undergrowth, which is like a forest to such a small creature. There is the threat of predators, and the dead bodies of other males are lying around. The male, like Theseus entering the Minotaur's labyrinth, finds the silken thread left by a female wherever she goes, and he follows it. It leads him face to face with her. But instead of being impressed by him, she tries to kill him. So he has to dance – maniacally, again and again – while avoiding her attacks. If he can't wear down her resistance and she kills him, his life will have amounted to nothing. But if he can impress her, they will mate, and his life's purpose will have been achieved.

In filming animals, there is a fundamental rule that the amount of equipment required is in inverse proportion to the size of the subject. The jumping spider is very small indeed. So the array of equipment included new miniature cameras, macro-camera-tracking systems and novel forms of lighting. And, of course, there were the ultra-HD cameras – except that, on a small scale, these become a liability. The bane of any macro-camera operator's life is trying to keep a minute subject in focus. So a camera with even shallower focus than normal was like a cruel joke – but one where the brilliance of the images made up for it.

The most important ingredient in realizing the stories is the one that never changes: the people, and the attitude and skills they bring. The problems they have to overcome are always the same, though in varying degrees. There are the technical challenges, and there is usually discomfort and frustration – long periods of intense concentration, maintaining a state of alertness and readiness to act, even when nothing seems to be happening.

But the most famous attribute of wildlife film crews, the one they are supposed to be most endowed with, is the one that's often misunderstood: patience. Undoubtedly a successful team needs the ability to wait and wait, often without moving, for hours, days or weeks – all for a moment of animal action which might last seconds. This sometimes seems to demand a kind of Zen state of mind. But patience is perhaps not the right word.

The key attribute of successful film crews is that they want each and every shot to be outstanding. Failure hurts them, but success is like an endorphin hit, and the best crews are addicted. If they have to sit motionless for hours on end in sub-zero temperatures, then they do. But not in a state of patience – it's an act of will. And when reading of the trials of *Life Story* cameraman Rolf Steinmann (page 276) and seeing footage of his attempts to film bonobos in the brutal Congo forest, the bitter-sweet reality of what he and the teams faced comes sharply into focus. These were indeed acts of will.

KNEELING WITH LANGURS
Barrie Britton waking up with the langurs to film early-morning social behaviour, in Jodhpur, India.

The perfectionist puffer

All the people, all the equipment, all the weather, and one little fish

The revelation that a fish was capable of creating an aesthetically beautiful, perfectly symmetrical construction was too good a story to miss. But the technological challenges this raised in filming a rare, camouflaged little fish at a depth of up to 25 metres (82 feet) were huge. In fact, it would require an underwater macro studio to be built and transported to the Japanese location.

It was a task for specialist underwater cameraman Hugh Miller. Months before the shoot, he started to build an underwater crane that could move the camera around the fish's wheel-like nest but wouldn't seize up when exposed to sand and sea water. Tests in the swimming pool went well. Lighting was always going to be a problem, as the nests are usually at a depth where sunlight is limited. But Hugh was confident that his revolutionary but distinctly Heath Robinson lighting contraption would do the job. Producer Miles Barton was less so.

For Hugh and his dive assistant Kat Brown to film behaviour for several hours at a time, they chose to use rebreather apparatus, which as well as enabling them to extend the filming time, had the advantage of not releasing bubbles and disturbing the pufferfish.

Eventually, in late June 2013, the crew and 700kg (1,543 pounds) of equipment arrived on the tropical Japanese island of Amami, and everything was assembled ready to go under water. Local expert Yoji Okata – Japan's foremost photographer of marine fish and the man who first solved the puzzle of the mystery circles – met up to discuss what he had found for them to film. The news wasn't encouraging. In the whole bay, there was only one male fish at a depth of less than 25 metres (82 feet), and he was already caring for his eggs and would not be building again for perhaps another week.

Despite being a septuagenarian, Yoji was extremely fit, and Hugh and Kat struggled to keep up with him as he led them out to meet the prospective star of the show. 'The biggest surprise', says Hugh, 'was how small the puffer was – half the size I was expecting from the photos I'd seen.' It was just 12cm (4.75 inches) long. Producer Miles Barton was also worried. 'Our hopes now rested on one little fish.' But while they waited to see if he would build another nest, they could at least test the equipment.

First the heavy quad – the four-legged tripod – was lowered over the side. Then the crane was, in sections, its weight borne by airbags attached like balloons to the frame. Once it was safely lowered 13 metres (42 feet) to the seabed, Hugh and Kat set about assembling it bit by bit. They resembled moonwalking astronauts as they strode to and fro across the seabed in apparent slow motion, with the emphasis on slow. 'We had to build it and unbuild it and build it again. Everything takes so much longer under water. Just to get it fully operational took more than two hours,' says Hugh.

Next it was the turn of the lighting rig – a triangular bracket to which a series of battery-operated lights in underwater housings was attached. It was towed into position and then tied to a breezeblock anchor, causing it to resemble a kite flying in the sky but with a halo of light underneath. Pleased with his 'little ray of underwater

STUDIO UNDER THE SEA
Hugh Miller setting up his undersea macro studio. He used a specially made four-legged quad to keep the heavy camera and its housing rock-steady on the seabed.

CENTRE OF ATTENTION
Hugh filming, staying motionless on the seabed, in this case for hours. The pufferfish was remarkably tolerant of the human and his contraption, allowing Hugh to capture the whole building process at close range.

A-FRAME SET-UP
Opposite page, left
Hugh setting the camera at the top of the A-frame to film the circle from above. When they had to abandon shooting, it was the last piece of equipment to be removed – which, in the end, saved the day.

VIEW OF PERFECTION
Opposite page, right
The view of the nest in all its symmetrical glory, thanks to Hugh's special A-frame.

The nest in comparison to the fish was huge, and he was darting all around it at high speed. Trying to follow him was hugely frustrating

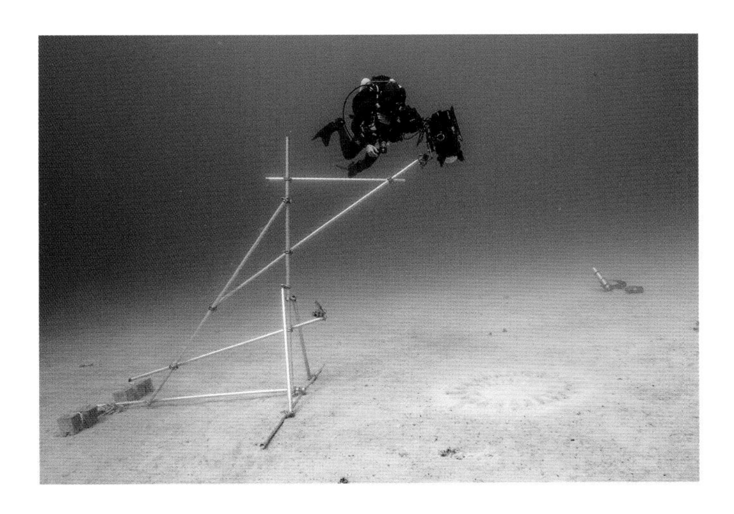

sunshine', Hugh then moved it into position beside the old nest, where the male was still fanning his eggs.

The fish wasn't at all fazed by the camera, the moving crane or even the lights. So all looked promising. And as he was still occupied with egg care, there was also time to build the 4-metre-high (13-foot) aluminium A-frame needed to hold the camera rigidly overhead to get the perfect topshot of the circle with the fish busy working away inside it. Everything was ready.

Miles, meanwhile, was lying awake worrying. 'Everything depended on this one male. What would happen if a hungry shark came by and snapped him up? Or if he just lost interest in the whole courtship business?'

He was right to worry. The next day the camera's control box sprang a leak and the delicate electronics flooded with sea water. Then Yoji returned with the news that the fish had started to build close to his old nest. It was a race against time. After a tense afternoon and evening drying out and reassembling the circuit board, Hugh finally got the box working again. The following morning filming began in earnest.

Hugh knelt behind the quad and tried to keep the pufferfish in frame. 'The nest in comparison to the fish was huge, and he was darting all around it at high speed. Trying to follow him was hugely frustrating.' Yoji had expected the fish to spend a week or so building – plenty of time for Hugh to get his eye in and film the behaviour in detail. But this little craftsman 'seemed to be turbocharged', says Hugh. 'On the first day, the nest was just a vaguely circular mess of disturbed sand, but by the second day, all the details of the ridges, furrows and peaks became clear.' And on the third day, Yoji returned from his afternoon dive and told the crew to be ready at dawn the following day. The speedy builder had just created a series of wavy lines in the centre of the circle, which normally meant that spawning would ensue early tomorrow.

The team was back in the water at 6am. 'It was still quite dark, but I could see the female swollen with eggs

STAR FISH
The tiny architect guarding his nest after the eggs have been laid. Keeping such a small subject in frame and in focus was a considerable challenge for Hugh.

as she came into the circle,' says Hugh. 'She had to run the gauntlet of the still aggressive male to get to the centre before he chased her away. Once she was in the middle, it seemed to trigger a change – as if now he was thinking, OK, its spawning time. He grabbed her face with his teeth, and they swam cheek to cheek. Then they spawned and parted. The whole process was repeated over and over until a bruise appeared on her cheek.'

Miles was delighted. Yoji's ability to find the little fish in the expanse of sandy seabed was impressive enough, but even more so was his ability to predict the very hour of spawning. 'With another 12 days filming ahead, I thought we'd be able to fill the gaps to get the complete sequence.' In the rush, what they hadn't achieved was the perfect topshot of the completed nest – to reveal the little fish's handiwork in all its glory. And as soon as spawning is over, the whole nest structure fades rapidly as the male switches from builder to father and fans the eggs. 'We just thought we needed to wait till he finished fanning his eggs and then started to build again,' says Miles.

But they hadn't banked on the typhoon. Working its way up the South China Sea, it was heading for Amami.

For four more days, while the sun shone and the sea was calm, the fish carried on fanning his eggs. Just as Yoji was expecting him to nest again, the wind began to rise, and the team was forced to remove the equipment and take shelter. All they could do was wait and watch as the surfers arrived to enjoy the big white waves.

A few days later the typhoon had passed, the sun was out and Yoji went back under water. Remarkably he found the same male fish. But every day the news was the same. He wasn't building. 'We learned that pufferfish are fussy,' says Miles, 'and he didn't like what the surge had done to the seabed.' Two weeks after spawning, the fish still showed no sign of building. So it was time to give up.

But as Hugh went in to remove the quad for the last time, he spotted the fish hard at work, building even faster than last time. Fortunately the metal A-frame was still in position, so Hugh was quickly able to attach the camera and get the missing top shot. 'If you'd told me in advance that we would have to rely on finding and filming just one little fish, I wouldn't have risked the shoot,' says Miles.

As for Hugh, he is just in awe of the fish himself. 'To shift that amount of sand and work that hard and fast day in and day out, you have to admire him. And the precision, symmetry and aesthetics of his work make him a class act.'

GETTING ON TOP OF IT
Hugh balancing himself above the A-frame. He had to keep his buoyancy neutral in order to stay perfectly still in the water, while both adjusting the camera settings and focusing on the circle below. He had only minutes to film before he had to return to the surface.

The chimps behaved, but the humans...

Bandits, hunters, explosions and a gold rush all added to a tricky shoot

The arid savannahs of southeast Senegal are home to a remarkable troop of chimpanzees – remarkable because of the behaviour they have devised for survival in this inhospitable region. Much of this behaviour had never been filmed before, and the team was to discover one reason why.

For the past ten years, anthropologist Jill Pruetz has studied the chimpanzees at the remote site on the Fongoli River, and so the team knew that they were used to seeing humans, but that, it turned out, was the only thing going for them. During the dry season, temperatures soar up to 45°C (113°F), and the chimps have to walk long distances every day in search of food and, more important, water, which they reach by digging shallow wells in the dry riverbed. They have also taken to using caves for siestas during the hottest part of the day. It's a tough season for the chimps, and it would prove to be even tougher for the team.

Accompanying director Emma Napper for the dry-season shoot was cameraman John Brown, whose job it was to follow these chimps as they searched for precious water, and Nick Turner, expert in low-light filming, who was tasked with rigging one of the cave-retreats with infrared (IR) cameras. But first they had to get there.

Just a few days before departure, reports began to filter through to the *Life Story* office of a bomb threat in the Senegalese capital Dakar and unrest around the borders of Senegal and neighbouring Mali, which was in the throes of a civil war. The level of danger needed to be checked out before the crew could be released to go, an essential process that took several days.

The delay was worrying – the crew needed to arrive at the height of the dry season, the only time the chimps dig for water and use the caves. As soon as they landed in Senegal, the team went straight into filming in a desperate effort to make up for lost time. Only three people were allowed to follow the troop in the field – John, accompanied by Jill's field assistants Michelle and Jonny. That meant John had to carry about 30kg (66 pounds) of filming kit, plus 6 litres (11 pints) of water every day, all the while wearing a surgical facemask to prevent possible transfer of human infections.

The team would get up at 3–4am, drive 40 minutes to the study site, then walk for up to an hour in the dark to get to the chimps before they woke up. The only way not to lose them was to stay with them all day until they settled down to nest for the night. Then the team had to trek back to the vehicle and drive back to base. If he was lucky, John would be in bed by 11pm – with an alarm set for 3am the following morning.

'The masks made it hard to breathe,' says John. 'The heat was brutal. And as most trees had shed their leaves, shade was in short supply and was typically occupied by the chimps, so you'd just slowly melt in the full glare of the sun.'

Meanwhile, Nick was trying to get to grips with the problem of rigging one of the ravine caves with his hi-tech cameras. After establishing the chimps were not in residence at the moment, Nick, Emma and two helpfully tall assistants, Zac and Massar, set about rigging the interior with a complex surveillance system of daylight

CHIMP AT WORK
A young chimp crafting a termite-fishing tool – a skill he's learnt – taking no notice of the human observers. Termites provide a year-round source of protein for the Fongoli community.

WATCHER AND WATCHED
Hearing the sound of the camera, the chimp stares at John for a moment before getting back to fishing for termites in the mounds. Most of the time, though, the chimps took no notice of their human followers.

BEDTIME
John filming the chimps making their night-nests at the end of the day. For John, though, getting to bed involved a one-hour trek and a drive, and he'd rise again at 3am to get back to the chimps before sunrise.

and infrared cameras and infrared lights. All this had to be powered and operated remotely from a hide 100 metres (328 feet) away in a dense thorn thicket. This is where Nick was destined to spend the next two weeks.

After six hours of exhausting work – using 500 metres (1,640 feet) of cables to link up six cameras, four monitors, ten car batteries and a control panel – everything was set for the chimps' return. After that, the daily routines for both crews settled into their patterns. While John tracked one troop through the scorching bush, Nick and Emma huddled inside the sweltering hide – 53°C (127°F) at its hottest – hoping that the chimps would make a visit.

It all started very encouragingly. On the first day, a beautiful male leopard wandered into the cave and curled up next to one of the cameras. It stayed for several hours before slinking off in the evening to hunt. This was pleasing because it was clearly not bothered at all by the lights and cameras, but it was also slightly worrying as, in the hide, it wasn't possible to know where the leopard was heading. But no chimps arrived that day or the next. This was not entirely unexpected – the chimps might well have been far away in their range. Then disaster struck.

Early one morning, Nick heard a number of loud explosions and volleys of gunshots. The local guide reassured him that this was probably local hunting and mining activity. But refugees from Mali were also pouring over the border, both to escape the war and to look for gold – activity that was starting to destroy the forest.

As far as filming the chimps in the caves, it turned out to be just as disastrous. The blasts were the result of

a local gold rush, mostly small-scale and unregulated, bringing with it all sorts of social and environmental problems. Every day brought more disturbance and noise, with the bush echoing to the sounds of gunfire and explosions and the general traffic of people, including bush-meat hunters.

As the days passed, it became clear that the chimps were not going to show. So reluctantly Emma and Nick derigged the cave. But far away, John was finally having some success as the troop made its move towards the dry bed of the Fongoli River. What was left of the water was stagnant and scummy, but the chimps knew to dig down to the water table to find fresh water. At the best places to dig wells, they would form patient queues waiting for cool, clean water to percolate through the gravel of the streambed into the hole. The first to drink were usually the dominant males and larger females, often sticking their heads right down into the hole for minutes at a time. Only when they had finished would the subordinate chimps take a turn. Capturing this remarkable digging behaviour was exactly what the team had been hoping for and helped compensate a little for the disappointment of the cave shoot.

Spending time so close to the chimps left a lasting impression on John. 'They were wonderful, truly fascinating creatures to spend time with. With every other species I've filmed – no matter how interesting or beautiful – there's a clear line between them and me. But with chimps that line is blurred, and you know you're sharing the environment with a consciousness that's very similar to your own.'

RED
RAVEN
WOODEN CAMERA

With chimps... you know you're sharing the environment with a consciousness that's very similar to your own

WATCHING THE WALK
Filming a chimp crossing one of the many open spaces in the savannah woodland. The harshness of their semi-desert home forced the chimps to walk huge distances to find food and water – 20 kilometres (12 miles) or more a day in the dry season. The humans would follow on foot, sinking into their rhythm and learning to anticipate their behaviour by the group's mood.

LOW-LEVEL ANGLE
Opposite page Michelle – an expert in the field who knew the chimps individually – observing. John is filming sitting cross-legged – his posture for low-level filming, using a short-legged tripod. When the chimps were on the ground, the objective was always to be at eye-level or below them to give the characters cinematic significance.

HOT AND SWEATY
Having to wear a mask all day in the heat felt suffocating, and sweat was a constant problem. The camera was state-of-the-art, as used by top-end feature films, but a heavy burden on the long daily treks.

Hunting the hunt

The show gets exciting when it goes to the dogs

One of the most ambitious of all the shoots was an attempt to film a pack of African wild dogs in hot pursuit of their prey across the plains. The very nature of the hunt makes it extremely challenging. Wild dogs can run at 50km/h (31mph) and may cover several kilometres. This is played out over rough terrain and often happens at night. Director Emma Napper and cameraman Jamie McPherson knew it was never going to be easy.

Even finding a possible location wasn't straightforward. For the previous year, the team had been in contact with scientists studying wild dog packs from Kenya to South Africa. The search had finally led us to a remote corner of Zambia, to Liuwa Plains National Park near the Angolan border.

The Sausage Tree Clan was large and had been studied for several years. Better still, it often hunted during the day. One of the dogs was radio-collared, and the alpha female had recently given birth to 12 pups, which meant there were lots of hungry mouths that needed feeding. But best of all, Liuwa, being part of the Zambezi floodplain, was as flat as a pancake *and* there were hardly any trees, just miles and miles of grass. It seemed perfect.

The plan was that Emma and Jamie would spend several weeks in the park, first covering the hunting dogs from the ground, then bringing in a helicopter to film the action from the air, to give the best chance of seeing the complex strategies the dogs used to find and then chase down their victims. Simple – or so they thought.

It didn't take long for the harsh realities to become apparent. Despite one of the dogs being radio-collared, finding them on the vast plain was a nightmare. The range of the transmitter was around 2km (a little over a mile), but the dogs rarely bed down in the same place twice and can easily roam more than 30–40km (18–24 miles) in a night. The light aircraft usually used for tracking was out of commission. So the team had to try to track the pack from the ground – night and day – and follow it whenever and wherever it moved. But even that plan didn't always work, and the dogs would disappear into the grass for days at a time.

One of the few predictable things about the African wild dog behaviour was the start of any hunt. 'From ground level it was relatively easy to see the start,' says Emma. 'The dogs have a rather elaborate, touching pre-hunt ritual, when the lead dog tries to galvanize the rest of the pack into action. They nuzzle and nip at each other, working themselves up into a frenzy before they head out.' But the difficulty was getting ahead of the hunt to film the end. There was also a downside.

In Liuwa, the dogs' main prey is wildebeest and lechwe (large, water-loving antelopes), both far heavier than the lightweight dogs. While big cats such as lions usually kill quickly by asphyxiation, the dogs simply begin to tear at a prey animal, disembowelling it while it stands. It was quite harrowing to watch – the victim eventually dies of shock and loss of blood. Even worse, packs of

GREETING, TRACKING, SPOTTING, ZOOMING
Top left The wild dogs working themselves into hunting mode – nipping and nuzzling, squealing and twittering. The behaviour is initiated by the alpha male, who with the alpha female, leads the hunt.
Top right Mutangh Dennis trying to locate the direction of movement of the radio-collared female, while Jamie scans the grass.
Bottom right Youngsters hidden in the grass, waiting for the adults to return from the hunt.
Bottom left Jamie with his tripod on the ground, following the fast-moving hunt.

ABV 839

PREPARING TO TRAVEL WITH THE DOGS
The charter plane dropping off the first load of filming gear at the remote campsite. There wasn't room for team, provisions and all the gear in the little plane, which meant a second two-and-a-half-hour flight.

OUTNUMBERED
A wildebeest waiting for the end after being chased to exhaustion by the wild dogs. Hyenas had tracked the hunt and were trying to take the prey. But this time they were outnumbered by the dogs.

hyenas often tracked the dogs and will fight them for their victims – often while the animal is still alive – turning it into a bloody tug-of-war.

Over the first few weeks of the shoot, Emma and Jamie managed to film the beginning and the aftermath of several hunts. But what happened in between – the nitty-gritty of the chase – was proving elusive.

'Once they got going, the dogs would just disappear over the horizon in a cloud of dust,' recalls Emma. The tall grass hid all sorts of obstacles – soft sand, termite mounds, water courses – that made it all but impossible to follow the chase at anything like the speed of the dogs, even in a 4x4. It was time to call in the helicopter – except that it turned out that there were no suitable helicopters available in Zambia. The last resort was to call on the help of experienced filming pilots from South Africa.

But just a few days before their arrival, all the Zambian Government clearances were declared null and void, and it became a race against time to wade through a bureaucratic minefield of new permits, permissions and paperwork to get the helicopter into the country in time. On top of that, with other commitments for both the crew and the helicopter, the window of opportunity was closing.

'It was so frustrating – the dogs were hunting pretty much every day, though at night, and everything was in place apart from the helicopter, grounded just a few hundred kilometres away,' says Emma. Then with just three days left before the helicopter had to be back in South Africa, the importation paperwork came through.

Now the pressure was really on. Instead of the planned ten aerial filming days, there were just two left. But as is so often the case, the filming gods suddenly smiled on them. On the first, rather speculative test flight, the helicopter rendezvoused with the researchers out on the plains to find the dogs already going through their pre-hunt rituals. Game on.

With Emma spotting from the front seat and Jamie in the back, surrounded by monitors and recorders and control panels, the helicopter took off as, right on cue, all 20 dogs – pups and adults – headed off across the plains.

'Initially, we stayed something like a kilometre away from the pack,' says Emma, 'because the powerful lens meant we could still pick out the individuals. But amazingly, the dogs totally ignored the helicopter, allowing us to follow them at closer range.'

After a few kilometres the dogs zeroed in on a small group of wildebeest. They stashed the young pups at a nearby waterhole and went into full hunting mode. Then, over 15 frenetic minutes, it all unfolded below the helicopter. Only from above is it possible to watch the action and see the strategies the dogs employ to single

out and bring down a single animal. 'They approach heads low, ears back, loping along in single file,' says Emma. The aim is to get the prey moving so the pack can assess which are vulnerable individuals. Once they have spotted a weakness, they accelerate to top speed.

'One dog initially takes up the lead and pushes hard, then drops back for a breather, letting others move through to maintain a relentless pressure,' she says.

But the wildebeest run hard and fast, kicking wildly if the dogs get close. They also try to find cover, heading towards other groups of wildebeest and zebras. The chase almost ground to a halt a couple of times when the dogs lost their quarry among the chaos of other animals. But they always managed to reorganize and switch pursuit to another wildebeest, one that must have appeared vulnerable in some way.

After tracking the hunt across several kilometres of rough terrain, the dogs finally latched on to a young wildebeest bull. 'Once they had locked on to their target, the whole pack came together,' says Emma, 'snapping and biting at its legs, tail and belly until they eventually dragged it to a halt.' When it was finally dead, the pack stood guard over the carcass while a subordinate dog set off to retrieve the pups from their hiding place and escort them back to the kill to feed.

Frustratingly, time ran out at the end of the day, and with the light about to go, the team was forced to leave the dogs to their feast. But they had seen the whole hunt play out beneath them, 'in all its bloody, exhausting, relentless violence' – a contrast to the cooperation and care the dogs show for each other and in particular for the pups.

'Despite all the blood and guts and the run-around they gave us,' says Emma, 'I have nothing but admiration for these truly incredible animals.'

FACE-OFF
Defending a hard-won kill, a dog faces off a hyena. Though spotted hyenas are both larger and more aggressive than African wild dogs, a large pack of dogs can keep the hyenas from stealing a kill while they swallow the meat on the spot. If there are pups, the dogs bring back undigested meat in their stomachs, to regurgitate to the waiting youngsters.

On a wild-goose chase

Planning, risk-taking and luck-making

A lot can change in 30 years. So when the team considered filming a sequence last attempted in the early 1980s, success was far from certain. Nonetheless, the potential for capturing the spectacle of barnacle goslings launching themselves off a huge cliff was considered worth the risk.

The reason no one had filmed in Greenland's Orsted Dal Valley since *Kingdom of the Ice Bear* was simple: there were no flights to a now-defunct research station, and the only way to access this remote valley was to mount a summer expedition. So a three-week trip was planned for June 2012 to explore the possibilities.

After travelling to Constable Point in Greenland, cameramen Ian McCarthy and Mateo Willis and their kit were flown by helicopter up the coast and dropped into the vast, glacial valley, much to the bemusement of the muskoxen. They had camping and climbing equipment, a rifle and a flare gun and just about enough training to deter a hungry polar bear if one should wander into the valley.

First impressions were encouraging as in the distance they could hear the sound of squabbling geese. Immediately after setting up camp, Ian and Mateo headed out to try to find a suitable filming location. After a long search they found a colony in a small canyon – almost certainly where the previous crew had filmed 30 years before.

Though it was midsummer, it was midsummer in the Arctic, and the rain, snow and high wind were only occasionally tempered by sun. And any let-up from the weather was counteracted by an increase in the clouds of mosquitoes. Worse than the weather was the terrain. Setting up to film by the imposing cliffs was hard going and dangerous, and the cameramen had to be extremely careful when positioning heavy filming equipment. There was a constant risk of falling or being hit by rock – virtually every step created a mini-landslide – and the consequences of injury in such a remote location could be severe.

It took an hour to scramble up the boulder fields and 45-degree scree-slopes just to reach the cliffs and then another hour to climb to the top, and numerous gear-ferrying trips had to be made each day. So it was decided that, rather than climbing daily to the clifftops, it would be safer to establish a camp for Mateo at the top. This gave him the best vantage point to film the nesting geese and more time to monitor the progress of the chicks. Meanwhile Ian stayed at the bottom, poised to capture the jumping behaviour when the radio call came from above.

The timing of the shoot was perfect. After the two days it took to set up camp, the first goslings began to hatch. Though Ian and Mateo could see about 20 nests in the gorge, only 6 offered good filming angles. Some were as high as 150 metres (492 feet) and some near the scree-slopes. There wouldn't be many opportunities to capture an impressive jump, and as it was midsummer, with 24 hours of daylight, a round-the-clock vigil was necessary.

About a day and a half after the chicks hatched, the parents grew agitated, and once they decided to fly from their nests, the goslings followed. At first, not knowing what to expect, both cameramen missed a lot of shots. The chicks would leap after their parents with little warning and were incredibly hard to pick out against the vast cliffs. All too often they were obscured by rock, leapt in un-filmable directions, bounced unpredictably or got stuck high up in crevasses. But after a while,

▶ **CLIFF SET-UP**
Setting up a boom. This enables the camera to be angled over the edge, ready for the jumps.

FILMING ON THE EDGE
Roped up and ready to shoot, Mateo operates the camera while watching output on the monitor, shading the screen with black cloth. Strong arms were needed.

Ian and Mateo started to recognize the behaviour that helped predict when and where the jumping would happen.

Filming the tiny chicks bouncing down distant cliffs in high winds makes for demanding long-lens work, but slowly and steadily the cameramen got their eye in. They collected jaw-dropping moments. But the ambition was to portray the event from the chick's perspective. This meant climbing down ledges. They even managed to position and operate a filming crane on a tiny precipice, so the cameras could film down over the nests. Each time the chicks threw themselves off the cliffs, Ian and Mateo were rooting for them. So it was distressing for them that so many of them fell awkwardly, hit rocks, got stuck or got lost. In fact, about a third never made it down alive.

Then about halfway through filming, Arctic foxes turned up, and the mood changed again. Any chicks found at the base by the foxes stood no chance of getting past them. A few families would make it down when the foxes weren't around, but more often than not the frantic calling between the chicks and parents was so loud that it alerted the foxes. In the face of such relentless carnage it was hard to believe that cliff-jumping was a viable strategy.

For the last few filming days, the plan was to follow the families as they made their way down the scree slope and across the valley to the river. But with the increasing presence of Arctic foxes, it seemed doubtful that any chicks would make it. Finally, on the second to last day, a family with three chicks completed an impressive jump. Ian and Mateo moved in closer, and though at first the geese were a little wary, they quickly accepted them, almost as if they understood that humans would deter the foxes. The geese waddled down from the slopes, right through the camp, and paddled onto the river. It was a moment of great triumph – both to have filmed it and to have regained hope that some chicks could still survive.

By the end of two weeks, the long hours, bad weather and treacherous terrain had taken their toll. Though Ian and Mateo were totally spent, they left in good spirits. The shoot – a gamble – had paid off. Though not enough had been filmed to portray the full drama of the events, when the production team saw the shots of the goslings jumping, they knew that a second trip the next year would clinch it, building on what Ian and Mateo had learnt on the first expedition. There can be few more dramatic examples of young animals enduring against all the odds, and it was only fitting that the filming should require similar effort.

WAITING FOR THE CALL
Mateo waiting, and waiting for developments at the high-rise barnacle goose nest. The helmet was needed for the regular rock falls. The rope was holding him safe on the cliff ledge, which also required a steep climb down. Two days after the eggs hatched, the goslings jumped.

THE DROP
Producer Tom Hugh-Jones scanning the ground below for signs of geese, goslings and Arctic foxes.

Outfoxed in the Arctic

Why wildlife film-makers have to factor in luck, or lack of it

Arctic foxes have been filmed many times before but mainly in summer, when the weather is manageable and the foxes can be relatively easily located, feeding pups at denning sites or hunting at bird colonies. But the *Life Story* team wanted to document youngsters over their first winter, hunting lemmings or scavenging from polar bears. They knew the shoot would be difficult because the foxes would be living a nomadic existence, but what they didn't expect was quite so much of the unexpected.

After exploring options, the team chose a Swedish site where scientists had been studying the Arctic foxes and where there were reasonable numbers following two years of lemming booms. Guided by a park ranger, cameraman Rolf Steinmann and director Sophie Lanfear began a four-week expedition. They identified three den sites and by the end of the first day had set up their Sami-style tepee on an otherwise featureless horizon without unnerving the foxes.

Soon Rolf had footage of young foxes coming and going from a den. A good start, and the weather was unusually mild. But just a few days into filming, it all changed with the arrival of extreme winds and snow. With minimal visibility, they had to rely entirely on GPS to navigate around the mountains. On the rare occasions when they encountered a fox, it would simply melt into the blizzard.

Admitting defeat, they retreated to their wind-battered tent to sit out the storm for three days. When a break in the weather finally arrived, the team immediately set up a hide close to the dens, where Rolf was to sleep. Between blizzards and gale-force winds, he was able to observe a pattern in the foxes' behaviour, but it wasn't one he wanted to see. The foxes would emerge from a den only in the late evening and then disappear into the night. Then, when moral was already at a low point, an even worse storm hit.

With 138km/h (85mph) gales tearing at the tepee walls and blowing in snow through every crack, they were forced to evacuate by skidoo. Totally reliant on GPS to navigate off the mountain, they lost their way and each other but finally all made it back to their cabin base.

When they eventually returned it was to find the camp buried. They dug out their equipment and set up the hide again, but with the foxes only active at night it was a losing battle. Then, on the penultimate morning, Rolf had a break. He spotted a fox hunting for lemmings. But as he reframed for a tight angle, the fox pounced, killed the lemming and vanished. Rolf was inconsolable and the shoot was over.

They decided to find a new location where, instead of following the foxes across the tundra, the foxes might come to them. Seal River Lodge, on Canada's Hudson Bay, fitted the bill. In its 30-year history, foxes had always been present in winter. It was also a known spot for observing interactions between foxes and polar bears.

To make sure of success, the team decided to stay for five weeks. The drawback was that without a vehicle, filming had to be on foot, with an armed guide to protect against polar bears. Almost as soon as they headed out, they encountered bears. Though their first encounters were disconcerting, Rolf and Sophie soon discovered that standing your ground and talking to a bear was enough to discourage it, and banging rocks together sent it running.

Aside from a few heart-stopping polar bear events and nights lit by the aurora borealis, the weeks passed

WIPE-OUT
The gale getting going. In the end, evacuation was the only option, abandoning the gear and navigating off the mountain with GPS.

slowly, without an Arctic fox sighting. The problem was too many foxes of the wrong kind. Red foxes had begun appearing around the lodge. Bigger and stronger than their white cousins, they are able to outcompete them, and as climate change causes the temperature to rise, red foxes are moving north and displacing them.

Even when the temperatures dropped below -20°C (-4°F), the red invasion showed no signs of retreat, and in five weeks the team only saw three Arctic foxes, fleetingly – because as soon as an Arctic fox caught sight of its larger rival it disappeared. Once again, Sophie and Rolf were forced to return to Bristol with 'you should have been here last year' ringing in their ears.

The decision was taken to give up and film another story. But the following winter, new reports of fox sightings began coming in from Arviat, a more northerly town on Hudson Bay that red foxes hadn't reached, and the Arctic foxes were foraging around human settlements in the day.

NOTHING TO DO BUT FILM THE BEARS
Sparring polar bears, killing time, waiting for Hudson Bay to freeze over. Rolf was killing time filming them, waiting in vain for the foxes.

FIRST ENCOUNTER
Justin Maguire filming a curious fox – a time for celebration, until the hunters came.

THE LEMMING LEAP
A young Arctic fox hunting. It had pinpointed the lemming moving under the snow and now had to break through the frozen crust, paws and nose first, and land straight on it. Success took a lot of practice.

Even more promising, lemming numbers were up. The hunt was on again. But now Rolf had another episode of unlucky timing. He'd already committed to another project, and so Justin Maguire took his place.

On arrival, Justin and Sophie found plenty of juvenile foxes, identified by their greyer winter coats. And a couple of days later, Justin filmed the first lemming hunt. It happened far away and not in good light. It wasn't the greatest shot, but it was a psychological triumph. Gradually they began filming more and more hunts. The behaviour was easy to predict. Once a fox heard something under the snow, it moved delicately, head cocked, ears swivelled. Pinpointing the prey, it would leap high into the air and plunge nose-first into the snow. The fact that so many jumps were unsuccessful, ending up with the youngsters stuck head down in the snow, made for humorous scenes. Not so amusing was that, unknown to Justin and Sophie, they weren't the only ones planning on shooting foxes.

The hunting season began on 1 November, and at C$25 per pelt, Arctic foxes are an easy source of cash. From then on, the foxes became impossible to film. Rather than giving up, Sophie and Justin decided to risk moving farther out on the frozen tundra. Locating foxes in bitter conditions wasn't easy, and after 16 days the snow was becoming too hard for the foxes to punch through.

As Hudson Bay froze over, the foxes followed the bears onto the ice to scavenge scraps from seal hunts. Finally, with ever-shortening days, plummeting temperatures and general exhaustion, it was time for the humans to leave.

It is rare to return to base having fulfilled the original brief, and this shoot was no exception. But the crew had a story – and a sense of achievement. All in all, they travelled more than 27,000km (17,000 miles) over two winters and had finally succeeded – a case of third-time lucky.

WINTER GAMES: SWEDEN V CANADA
Opposite, top left Rolf in his 24-hour snow-hide.
Top right Sophie digging out the buried tepee in Sweden. It took many hours.
Bottom left Justin far out on Hudson Bay ice, following the foxes following the polar bears.
Bottom right Sophie and Thomas the guide in Canada. A skidoo pulled the gear on the sled.

Bonobos hide from all their cousins

Gorillas compete, chimps attack and humans come with their cameras

The bonobo is the least studied and filmed of all the great apes. There are only a couple of places where they can be approached in the wild, both deep in the Congo rainforest. After eight months of careful planning, researcher Theo Webb, cameraman Rolf Steinmann and field assistant Ed Anderson flew to Kinshasa to meet Gottfried Hohmann, who has studied bonobos for 20 years. His specific interest in the bonobo is its close relationship to us and the glimpse it may offer into the lives of our proto-human ancestors.

The team then bought supplies for the eight weeks of filming that lay ahead, including a lot of chilli sauce that would prove a godsend, and had their last meal out, fried caterpillars – 'fishy and crunchy', according to Theo.

The next leg of the trip was a flight to an airstrip in the forest. 'The whole village had trekked there to welcome us,' says Theo. Fifty porters then carried their supplies the 5km (3 miles) to the village where they had their last taste of meat on the trip, red river hog, and were introduced to manioc, the staple carbohydrate food. A parsnip-like root vegetable, it can be fried or boiled and then fermented. The last form is known as quanga and, according to Theo, 'is absolutely disgusting', but it would be essential fuel for the exertions that were to come.

The following day they made the 25km (16-mile) walk to the LuiKotale research camp that would be their home for the next two months. Gottfried, who set up the camp in 1992, advised the team on etiquette. 'Make the bonobos aware of your presence. Rip leaves and talk as you approach so they know you are coming. You must show your faces; don't try to hide in any way. They need to see you among the familiar faces [the scientists and trackers]. Then they'll start to recognize you. And don't stare – that's a threat.' The team also had to wear surgical masks to avoid passing on infections and were advised not to sit directly beneath the apes to avoid being peed on or worse.

After his first day in the forest, Rolf was feeling positive. 'I knew it was going to be tough, especially in the humidity, but we had lots of time to find the animals.' His optimism, however, didn't last. It took three days before he even saw a bonobo.

A typical day meant getting up before dawn and, with a 25kg (55-pound) rucksack, walking anything from 5km to 20km (3–12 miles) to where the bonobos had

▲ **THE LAST MEAL OUT**
Fried caterpillars – crisp and fishy. It was a protein-rich meal the team would often think about longingly.

▶ **BASE CAMP**
LuiKotale research camp in the heart of the Congo rainforest, complete with solar panels – home for two months.

ON THE BONOBO TRAIL
The team following the bonobos searching for a new fruiting tree. The daily trek was often more than 25km (16 miles), carrying all the camera equipment plus water. An encounter with army ants often meant stripping off – an unpleasant daily routine, sometimes more common than sightings of the bonobos.

spent the night. Then they would sit down and wait until the sun came up and there was enough light to film. They could hear the bonobos high in the trees but couldn't see them. Eventually the animals would come down and head off to find a fruiting tree. Only in the brief time they spent on the ground did Rolf have a chance of filming them.

'The group would stay in a tree all morning,' says Rolf, 'then walk on the ground for a short time before leading us into the path of yet another column of army ants.' Theo remembers the ants well. 'You had to run to get away from them, then strip off your clothes to remove them one by one. But the most disturbing thing was the way they seemed to coordinate their attack. Only when they were all over you and already deep into your clothing did they bite, all at once.'

Other hardships included the sweat bees that formed clouds around their heads, attempting to enter any available orifice, and the thousands of termites that would invade any packs left on the ground. And then there were the honeybees that were attracted to the sweat on the rucksacks. They would cluster around the straps and sting as soon as a pack was hoisted up.

After weeks of failing to film and a barrage of insect attacks, Theo was worried. 'Initially Ed and Rolf were really buoyed up by the amazing experience of meeting the bonobos, but after three weeks walking an average 25km (16 miles) a day without getting a single shot they were in despair.'

'We were constantly setting up the camera, trying a shot, failing and having to move on,' says Rolf, 'all day every day.' On one rare occasion when they were in a good position to film the bonobos making a nest, the camera packed up and they had to leave the animals in a perfect

filming position and walk the 5km (3 miles) back to camp. 'Some days you just wanted to forget as quickly as possible,' says Rolf.

And the food didn't help. A daily diet of red beans and quanga manioc – garnished if they were lucky, with tinned sardines – was taking its toll. Theo, who was in charge of preparing lunch, was getting a lot of stick, despite his best efforts to liven the food with chilli sauce. So when the camera broke down and an airlift of another camera had to be arranged, he spotted an opportunity. He ordered something guaranteed to be tasty – a Camembert cheese. 'I could smell it before it arrived in camp. It was runny, but the camp had a feast that night, and we ate every morsel.'

FOOT REST
Field assistant Ed (left) and cameraman Rolf (right) collapsed after a rare return to camp in daylight. The worst pain was often in the feet, which became blistered and infected with fungus after being encased all day in wet boots.

The struggle went on. 'Ninety per cent of what you see you can't film,' says Rolf, 'because the vegetation gets in the way. So you really have to learn to deal with the frustration – become a Zen master and suppress your emotions. Otherwise you face a high risk of going insane.'

Eventually, after six weeks, their luck began to turn, and on a couple of occasions they were able to film a group of bonobos grooming each other on a fallen log. But what they really wanted to capture was a scene that might resemble the lives of our earliest ancestors. So Rolf and Ed were very excited when the group finally headed for the swamp, which was a little more open than the rest of the forest.

The bonobos picked their way through the water, walking on two legs, plucking and chewing on lily stalks (believed to be an important source of minerals). Mothers holding their young on their shoulders could have been human mothers travelling on foot. And Rolf had a reasonable view of them for a full 20 minutes.

He was ecstatic. 'We thought it was impossible to film but we filmed it. You can't believe how difficult it was. I was filming through this one little hole in a bush that formed a natural hide for me. It was behaviour that has never been filmed before. So it was wonderful for us.'

The result was a truly privileged portrait of some of our closest yet least-known relatives.

THE REWARD
A very young bonobo being cradled by its adoring mother – a sight glimpsed through vegetation but a reward well worth waiting for. Vegetation was one of the biggest problems that Rolf faced, along with the lack of light in the forest and the fact that the bonobos spent a lot of time high up in the trees. So every sighting was a cause for celebration.

Index

Page numbers in *italic* refer to illustrations

Acknowledgements

This book is published to accompany the TV series *Life Story*, first broadcast on BBC1 in 2014
Executive producer: Michael Gunton
Series producer: Rupert Barrington

1 3 5 7 9 10 8 6 4 2
Published in 2014 by BBC Books,
an imprint of Ebury Publishing
A Random House Group Company

The Random House Group Ltd Reg. No. 954009

Addresses for companies within the Random House Group can be found at www.randomhouse.co.uk

A CIP catalogue record for this book is available from the British Library

ISBN 978 1 849 90664 7

The Random House Group Limited supports the Forest Stewardship Council® (FSC®), the leading international forest-certification organisation. Our books carrying the FSC label are printed on FSC®-certified paper. FSC is the only forest-certification scheme supported by the leading environmental organizations, including Greenpeace. Our paper procurement policy can be found at www.randomhouse.co.uk/environment

Commissioning editor: Muna Reyal
Project editor: Rosamund Kidman Cox
Designer: Bobby Birchall, Bobby&Co Design
Picture researcher: Laura Barwick

Colour origination by AltaImage, London
Printed and bound in Italy by Printer Trento

To buy books by your favourite authors and register for offers visit www.randomhouse.co.uk

For as long as wildlife documentaries are made, all of us involved in their production will continue to owe a deep debt to the scientists, field experts and local support who make it possible for our crews to be in the best places at the best times, informed in detail about what to expect.

Over four years the tentacles of the *Life Story* production team have reached right around the globe. From the Arctic Circle to Africa and down to the Antarctic; from the Indian subcontinent to the remote Gobi Desert and beyond into the Far East; through the Americas to the multiple corners of Australasia. In every part of the world, experts have selflessly given both their time and the hard-won fruits of their careers: their intimate understanding of how their animal subjects live their lives. Why should they do this? There is not much money on offer and no fame. They do it because they are deeply fascinated by the creatures that they study, and they approach their work with an integrity and openness that is rare in many other spheres of life. Their passion is what made the TV series and this book possible.

They are of course simply too numerous to acknowledge individually and we thank them all equally. But to give a sense of the scale and generosity of their support, I mention a few. Scientist Jill Pruetz has studied a population of chimps on the edge of the Sahara in Senegal for a decade, achieving a detailed understanding of each individual in the troop. Her knowledge and support enabled our crew to record remarkable footage of the chimps' culture of spear-hunting, which she originally discovered there. Japanese photographer Yoji Okata discovered both the sand 'crop circles' off the coast of Japan and the pufferfish that makes them. It would have been impossible for our crew to film this phenomenon without his help and knowledge. Scientist Konstantin Rogovin allowed our team to join a Russian expedition that took them deep into the Gobi Desert. There he located the extraordinary long-eared jerboa, which he had studied 17 years previously.

We also owe grateful thanks to the team behind the production of this book – a complex and unsung task. Muna Reyal commissioned it and Albert DePetrillo took on her role when she moved to new pastures, assisted by Kate Fox. Bobby Birchall is responsible for the smart design. Laura Barwick has located novel pictures to illustrate the stories. Roz Kidman Cox has once again proved a wise and inspiring editor.

Production team
Rupert Barrington
Miles Barton
Alison Brown-Humes
John Bryans
Tom Crowley
Carlee Davis
Nick Easton
Katie Ellis
Joseph Fenton
Sandra Forbes
Ian Gray
Michael Gunton
Craig Haywood
Karen Hooper
Tom Hugh-Jones
Ellen Husain
Nadege Laici
Alex Lanchester
Sophie Lanfear
Lannah McAdam
Emma Napper
Anuschka Schofield
Nick Smith-Baker
Theo Webb
Lucy Wells
Loulla Wheeler
Matthew Wright

Camera team
Matt Aeberhard
John Aitchison
Guy Alexander
Doug Anderson
Luke Barnett
Barrie Britton
John Brown
Rod Clarke
Martyn Colbeck
Tom Crowley
Sophie Darlington
Rob Drewett
Dawson Dunning
Tom Fitz
Kevin Flay
Dave Griffiths
Ben Grover
Jeff Hogan
Richard Jones
Mark Lamble
Emilien Leonhardt
Ian McCarthy
Alastair MacEwen
Mark MacEwen
Jamie McPherson
Justin Maguire
Hugh Miller
Roger Munns
Peter Nearhos
Didier Noirot
Mark Payne-Gill
David Reichart
John Shier
Mark Smith
Rolf Steinmann
Paul Stewart
Toby Strong
Gavin Thurston
Jeff Turner
Nick Turner
Simon Werry
Mateo Willis
Kim Wolhuter
Richard Wollocombe

With special thanks to
Thomas Alikashwa
Ed Anderson
Anders Angerbjorn
Shinsuke Asaba
Leesa Baker
Tudevvaanchig Battulga
Matthew Becker
Ian Bell
Jami Belt
Brett Benz
Hakan Berglund
Emily Best
Kelly Boyer
Kat Brown
Rob Byatt
Rachel Cartwright
Anil Chhangani
Nyamtseren Choinzon
Tim Clutton-Brock
Andy Collins
Matthew D'Avella
Kleber Del Claro
Cassandra Denne
Mutangh Dennis
Stephanie Doucet
Egil Droge
Andy Dunstan
Ewan Edwards
Vicki Fishlock
Madeline Girard
Tyler Goertzen
Petr Gunin
Philippe Henry
Toyo Hirohashi
Gottfried Hohmann
Martin How
Crissy Huffard
Michael Huffman
Samuel Jaffe
Heather Jooste
Dondo Kante
Bakary Keita
Jeroen Koorevaar
Jose Lachat
Stuart Lamble
Andy McPherson
Julio Madriz
Terence Mangold
Marta Manser
Ray Mendez
Artur Miguel Vitorino
Joseph Mobley
Karina Moreton
Cynthia Moss
Sammy Munene
Norah Njiraini
Chris Likezo Numwa
Yoji Okata
Hans-Dieter Oschadleus
Jurgen Otto
Christina Painting
Chip Payne
Doug Perrine
Jerome Poncet
Jill Pruetz
Simon Robson
Konstantin Rogovin
Randi Rotjan
Tara Ryan
Georgy Ryurikov
Santos
Katito Sayialel
Robert Sayialel
Soila Sayialel
Stefan Schuster
Gautam Sharma
Digpal Singh
Rick Sinnott
Claire Spottiswoode
John Staniland
Alexey Surov
Michel Tama Sadiakhou
Glen Thelfro
Everton Tizo-Pedroso
Miquel Torrents Tico
Amy Venema
David Wagner
Tom Walker
Jody Weir

Post production
Linda Castillo
Miles Hall
Janne Harrowing
Rupert Howe
Esta Porter

Music
BBC National Orchestra of Wales
London Session Orchestra
Murray Gold

Film editors
Nigel Buck
Darren Flaxstone
Angela Maddick
Andrew Mort
Dave Pearce

Dubbing editor
Paul Cowgill

Dubbing mixer
Graham Wild

Colourist
Adam Inglis

Online editor
Tim Bolt

Graphic Design
Mick Connaire

Discovery Channel
John Cavanagh
Robert Zakin

France 5
Thierry Mino
Perrine Poubeau

The Open University
Julia Burrows
David Robinson
Janet Sumner
Vicky Taylor

Picture credits

1 Oldrich Mikulica; **2–5** Federico Veronesi; **6–7** Sophie Lanfear; **8** Kevin Flay

1 FIRST STEPS

10–11 Grégoire Bouguereau/vieimages.com; **13** J-L Klein & M-L Hubert/FLPA; **15** Stefano Unterthiner; **16 *top left*** Stephen J Krasemann/SPL; **16 *top right*** Ingo Arndt/naturepl.com; **16 *bottom left*** Stan Malcolm; **16 *bottom right*** Darlyne Murawski; **19 *top left*** BBC; **19 *top right*** Rob Byatt; **19 *bottom left*** BBC; **19 *bottom right*** Chien Lee/Minden/FLPA; **20** BBC; **21** Tom Hugh-Jones; **22–3** Ian McCarthy; **24–5** BBC; **26** Michel Laplace-Toulouse/Biosphoto/FLPA; **28–9** Anup Shah/shahrogersphotography.com; **30–1** Juan Carlos Munoz/naturepl.com; **33** Brandon Cole; **34–7** Cesere Brothers Photography/NMFS Permit #10018; **38** Andy Rouse/naturepl.com; **40** Suzi Eszterhas/Minden/FLPA; **41** Tony Heald/naturepl.com; **42–3** Daniel Rosengren; **44** Solvin Zankl/naturepl.com; **45** Shem Compion/shemimages.com; **46** Theo Webb; **47** Robin Hoskyns

2 GROWING UP

48–9 Will Burrard-Lucas/burrard-lucas.com; **51** Luciano Candisani; **53** Mitsuaki Iwago/Minden/FLPA; **54** Paul Nicklen/National Geographic Creative; **55** Sophie Lanfear; **56** Matthias Breiter/Minden/FLPA; **57** Steven Kazlowski/naturepl.com; **58–9** Eric Baccega/naturepl.com; **60** Frans Lanting/FLPA; **61** Bill Curtsinger/National Geographic Creative; **62** Doug Perrine/naturepl.com; **63** Jessica Farrer; **64** Theo Webb; **65** Tom Hugh-Jones; **66** Andrew Parkinson/naturepl.com; **67** Danny Green/naturepl.com; **68–9** Theo Webb; **70** Frans Lanting/FLPA; **71** Pierre Lobel; **72** Frans Lanting/FLPA; **73** Laurent Demongin; **74** Jurgen Freund/naturepl.com; **76–8** BBC; **79** Barry Hatton; **80–1** Tim Laman/naturepl.com; **82** M & P Fogden/fogdenphotos.com; **84** BBC

3 HOME

86–7 Tim Laman/National Geographic Creative; **89** Vincent Munier; **91** Jurgen Freund/naturepl.com; **93** Shattil & Rozinski/naturepl.com; **94** BBC; **95** Shattil & Rozinski/naturepl.com; **96–7** BBC; **98–9** Alexander Safonov; **101** Sumio Harada/Minden/FLPA; **102** Skip Brown/National Geographic Creative; **103–5** Sumio Harada/Minden/FLPA; **106** Adrian Bailey/baileyimages.com; **107** Shem Compion/shemimages.com; **108–9** BBC; **110–1** Will Burrard-Lucas/burrard-lucas.com; **113** John Brown; **114–5** BBC; **116–7** Jiri Slama; **118** Art Wolfe/artwolfe.com; **119** Ingo Arndt/Minden/FLPA; **120** Mark Moffett/Minden/FLPA; **121–3** BBC

4 POWER

124–5 John L Dengler/denglerimages.com; **127** Doug Perrine/naturepl.com; **129** Erlend Haarberg/naturepl.com; **130–3** John L Dengler/denglerimages.com; **134** Mike Potts/naturepl.com; **135–7** John L Dengler/denglerimages.com; **139** Donald M Jones/Minden/FLPA; **140 *bottom left*** Jeff Vanuga/naturepl.com; **140–5** BBC; **146** richarddutoit.com; **147** Beverly Joubert/National Geographic Creative; **148–50** brendoncremer.com; **151–3** Pete Oxford/Minden/FLPA; **154** Frans Lanting/FLPA; **155** Emma Napper; **157** BBC; **158–9** Frans Lanting/FLPA; **160–1** John Brown

5 COURTSHIP

162–3 Steve Race; **165** Andy Rouse/naturepl.com; **167** Tim Laman/National Geographic Stock/naturepl.com; **168–71** Christina Painting; **172** Marcel Gubern; **174–5** Gilbert Woolley/Scubazoo; **176–7** Tanya Detto; **178–9** BBC; **180** Paul Stewart; **182 *left*** Otto Plantema; **182 *right*** BBC; **183–4** BBC; **185** Kat Brown; **186–91** BBC; **193** Miles Barton; **194** Yukihiro Fukuda/naturepl.com; **195** BBC; **196–7** Yukihiro Fukuda/naturepl.com; **198–201** Jurgen Otto

6 PARENTHOOD

202–3 Christian Ziegler; **205** Thomas Dressler/ardea.com; **207** Patrick J Endres/AlaskaPhotoGraphics.com; **208–11** Gary Bell/Oceanwide images.com; **212** BBC; **213** Miles Barton; **214** Alex Lanchester; **215–6 *top left*** Mark Payne-Gill; **216 *top right*** Claire Spottiswoode; **216 *bottom left*** Alex Lanchester; **216 *bottom right*** Claire Spottiswoode; **219** Bernard Castelein/naturepl.com; **220–3** Stefano Unterthiner; **224–7** Darlene Boucher; **229–31** Christian Ziegler; **232–3** Theo Webb; **234** Denis-Huot/naturepl.com; **235** Anup Shah/shahrogersphotography.com; **236–7** Michael Nichols/National Geographic Creative

7 *LIFE STORY'S* OWN STORY

238–9 Sophie Lanfear; **241** Rolf Steinmann; **242** Alex Lanchester; **243** Corinne Chevallier; **244** Alex Lanchester; **245** Jurgen Otto; **246** Emma Napper; **247** Alex Lanchester; **249–50** Kat Brown; **252** Tom Crowley; **252** Yoji Okata; **253** Tom Crowley; **255–6** Emma Napper; **257** Ian McCarthy; **258–9 *top*** Emma Napper; **258–9 *bottom*** Ian McCarthy; **261–2** Emma Napper; **263** Alex Lanchester; **264–5** Emma Napper; **267** Mateo Willis; **268** Zac Poulton; **269 *left*** Mateo Willis; **269 *right*** Zac Poulton; **271** Rolf Steinmann; **272–3** Sophie Lanfear; **274** Steven Kazlowski/naturepl.com; **275 *top left, bottom left*** Sophie Lanfear; **275 *bottom right*** Rolf Steinmann; **276** Theo Webb; **277–8** Edward Anderson; **279–81** Theo Webb

Endpaper *front* Frans Lanting/FLPA; **Endpaper *back*** Cesere Brothers/NMFS Permit #10018